JET GIRL

JET GIRL

MY LIFE IN WAR, PEACE, AND THE COCKPIT OF THE WORLD'S MOST LETHAL AIRCRAFT, THE F/A-18 SUPER HORNET

CAROLINE JOHNSON

WITH HOF WILLIAMS

St. Martin's Press

New York

This book is a memoir. It reflects the author's present recollections of experiences over time. Some names and details have been changed to protect the identity of those involved. Some events have also been compressed, and the dialogue has been re-created based on the author's memory.

First published in the United States by St. Martin's Press, an imprint of St. Martin's Publishing Group

JET GIRL. Copyright © 2019 by Caroline Johnson. All rights reserved.
Printed in the United States of America. For information, address St. Martin's Publishing Group, 120 Broadway, New York, NY 10271.

www.stmartins.com

Designed by Devan Norman

The Library of Congress Cataloging-in-Publication Data is available upon request.

ISBN 978-1-250-13929-0 (hardcover)
ISBN 978-1-250-13930-6 (ebook)

Our books may be purchased in bulk for promotional, educational, or business use. Please contact your local bookseller or the Macmillan Corporate and Premium Sales Department at 1-800-221-7945, extension 5442, or by email at MacmillanSpecialMarkets@macmillan.com.

First Edition: November 2019

10 9 8 7 6 5 4 3 2 1

To the trailblazers who paved the way;
to the warriors currently in harm's way;
to future generations of men and women who will
raise their right hand;
and to all who encourage, mentor, and support us:
This one's for you.

JET GIRL

PROLOGUE

★

Shoes were off and my feet, in polka-dot socks, were on my desk next to a stack of letters from my ninety-six-year-old Grammy. Country music played on low in my earbuds. I blew strands of hair from my eyes and looked up from working on my master's to gaze at Ryan Gosling, who stared back at me from multiple angles of the many posters we had hung on our walls, unwaveringly sensitive and understanding. My full-time job, as usual, had kept me up the night before, but I'd pulled myself out of bed early to study. I felt groggy and missed my family, the comforts of home, and a perennial on-again-off-again boyfriend who I'd maybe never see again. Like most mid-twenty-year-old students balancing a job, academics, and a love life (or lack of one), I was lonely, overworked, and just trying to get by. In the hallway outside my room I could hear what amounted to a fire drill. I was beginning to get annoyed. I heard five life-altering words boom over the intercom, "Launch the alert 30 fighter."

I yanked out my earbuds, skin tightening and heart beating so hard the concussions reverberated in my teeth. The voice repeated the order, "This is the TAO. Launch the alert 30 fighter. I say again, launch the alert 30 fighter."

This command from the Tactical Action Officer, or TAO, would, in effect, send me flying—no, screaming—at nearly the speed of sound, armed to the teeth, into an international incident with one of our country's most volatile adversaries.

My dorm room for the past four months, "the Sharktank," housed the six fixed-wing female aviators aboard the Nimitz-class aircraft carrier USS *George H.W. Bush.* The *Bush,* a nuclear-powered, floating steel fortress known as "Mother," steamed at the head of a strike group. Imagine a dozen boats skimming across the glittering waters of the North Arabian Gulf in a V-shaped formation, like a necklace with the jewels drawing white wakes in the brilliant blue water. Don't let the pretty picture fool you. These warships bristle with enough weaponry to turn an area the size of Texas into a smoking pile of uninhabitable ash. And what you cannot see is doubly dangerous. Down in the depths of the ocean, armed submarines plow steadily ahead, sonars on, silent and battle-ready.

The purpose of the surface formation, the submarines below, and the seventy-five hundred highly trained Sailors manning the vessels was singular: guard the crown jewel traveling at the apex of the V-formation of the necklace. In other words, protect Mother, the most lethal and versatile weapon the US can drive out onto the world's stage. Of the nearly eight thousand people working in Strike Group Two, only three women flew the most feared plane of all, the most technologically advanced, the baddest of the bad—the $80 million F/A-18 Super Hornet. One of those three lucky girls was me, Caroline Johnson, full-time grad student, Naval Flight Officer, and Weapons System Officer. Jet Girl.

I jolted out of my chair and slammed my polka dot–clad feet into the boots directly next to my desk. Officers on alert are supposed to stay in a complete uniform at all times, but our steel-toed boots were waterproof, and I'd worn them more than fifteen hours a day, every day, for the previous four months in a climate so hot and humid you could drink the air. Rather than let my feet rot, I'd broken a rule. No time to worry about that.

Throwing the Sharktank door open, I sprinted down the hallway up the narrow ladder wells, taking two steps at a time.

"Alert 30 fighter, vector 330," the TAO said, his voice growing anxious over the intercom. Vector 330 was the heading where we would find whatever was stirring the hornet's nest. I knew I would need to recall it later, as I could already feel the boat turning to a different heading. The four engine shafts propelling the ninety-thousand-ton carrier churned at full bore, pointing Mother's bow into the wind in order to generate twenty-five knots of airflow across the bow. The wind over the deck provided a crucial boost for our fully loaded fighter until our two afterburners could power the plane into the sky.

Because of the fire drill, most of the ship's doors—monstrously heavy hunks of metal sealed watertight—were all closed as if we were under torpedo attack. I heaved open the massive door at the end of the first hallway, running into a crew of firemen performing a drill in the passageway. Under normal circumstances, I might have snickered at the six Sailors in full firefighting gear, pantomiming the act of extinguishing a nonexistent fire, but right then, they were blocking my passageway.

"Make a hole! Out of the way!" I sprinted toward them, but they continued their charades.

I braced myself, lowered my shoulder, and bodychecked the first dude, sending him smashing into the wall. His fire-team leader turned to me and snapped, "Someone's in a hurry."

"Get out of my way!" I pointed to the intercom overhead. "There's real-world shit going on!"

He looked up, blinking, confused, and then nodded to his men. "Move!"

I ran full-out down the hallway, knowing I needed to get from my stateroom in the bow of the *Bush* to the Ready Room all the way in the stern, about two hundred yards—or two football fields—away. My lungs burning and flight suit soaked, I hurdled the fifteen-inch braces that dotted the hallway every twenty feet. By yard one hundred and fifty, I was losing steam when the Carrier Air Group commander, or CAG, the second most powerful person in the strike group, ducked out of his office, clipboard pumping in hand.

"Keep going, Dutch!" he barked hoarsely. "Iran is forty miles out!"

A group of aviators waited for me in the Ready Room. "Go, go, go!" one of them said, shoving my helmet bag, complete with the encrypted codes and data for the jet's weapons systems, into my hand. "Crocket is waiting for you."

Crocket, aka Corn Rocket, was my pilot that day, and when I pushed into the paraloft, he was already halfway into his G-suit, deep into his pregame mindset.

"CAG told me an Iranian F-4 is inbound within our vital area," I said and he nodded. We both understood a line in the sand had been crossed. The Iranians flatly knew better than to fly so close to us, and our job was to go intercept that plane.

"See you up there." He winked and hurried up to the flight deck. Pilots and their Weapons Systems Officers, or WSOs, have a unique working relationship, a bond built on respect and trust. The respect is based on each other's skills, reputation in the air, and by proving time and time again that you can not only do the job, but do it flawlessly. I had tremendous confidence in Crocket's skills, as I knew he did in mine.

Trust is something different from respect. It is respect in a nosedive. Crocket would be flying what amounted to a fully loaded bomb at five hundred miles per hour and then landing it on a steel cork. I would be his eyes, ears, and guns as he did so. Even the most hardass, war-tested Navy SEALs think what we do is crazy. And when *those* guys think what you do is crazy, that's saying something.

Eschewing the normally regimented—almost ceremonial—calm and unhurried preflight, I yanked on my $150,000 joint helmet-mounted cueing system and, leaving my harness half-zipped, ran out behind Crocket. The colossal, watertight door separating the dark innards of the ship from the bright light of the North Arabian Gulf was so smothered in grease that it looked like it had been dunked in motor oil. No civilian would have dared touch the handle, but I heaved the large lever up and laid my shoulder into it. The hinge creaked loudly as I stepped onto the stairs, my senses blinded by the Middle Eastern sun and wall of hundred-degree heat.

I climbed the greasy steps, hoping that if a plane was parked above the staircase, it didn't have bombs under its wings. My joint helmet, which weighed about ten pounds, made it nearly impossible to look up. Forget Starbucks. Try head-butting a live warhead to wake yourself up in the morning.

Luckily, the only thing I crashed into was the overpowering reek of bacon as the vents from the ship's kitchen breathed the heavy blue smoke of institutional breakfast meats mingled with the scent of grease. Yum. *Now where is my jet?*

There, on the warship's massive deck, all the world's premier Tailhook aircraft were lined up, combat-ready, bathed in slanting morning light. After four months on a combat cruise, the jets were pretty beat up. Sandstorms in Afghanistan literally scratched off patches of paint so that the solid gray looked like gray and brown camo, all splattered with globs of grease.

The morning dew mixed with the layers of oil on the flight deck and I slid in my boots, weaving toward the Navy's version of priority parking—which, surprisingly, was *not* easy to find. Onboard the carrier, there is simply not enough room to park all the planes, so they are positioned inches apart, wings folded up in a puzzling maze of steel and explosives.

Each jet is marked with a unique number and the name and call sign of an aviator in the squadron. To ensure the fleet is rotated, for daily flying, every pilot–WSO pair is given whatever plane is currently "up" or ready to fly, and that day it just so happened that my jet, #210, was pulled up behind the catapult. I got a little warm feeling when I saw it:

LT CAROLINE JOHNSON
DUTCH

Since we'd already preflighted our plane before the alert, I knew everything was ready to go, but it still felt strange not to perform the usual exterior checklist: circling the plane, checking one last time to make sure the beast was secure and ready. I climbed into my seat behind Crocket and lowered the canopy as my free hand went into autopilot mode, strapping my harness into the ejection seat. The massive glass canopy met the rail, sliding forward and locking to pressurize the cabin. My ears popped as my fingers—nails painted with princess-pink polish—flew through checks, pressing buttons, flipping switches, and lightly tapping data into the touchscreen.

Outside, an eighteen-year-old plane captain in a brown shirt began to detach the six chains that secured our aircraft to the deck of the moving boat. I clipped my mask to my helmet, feeling the two familiar clicks, and flipped on the oxygen. I adjusted my French braid and gave it a soft tug—a bit of Jet Girl good luck. The braid was also a reminder that as a woman I was willing to die for my country that lets me fly an $80 million plane while women in other nations aren't allowed to drive a car.

The flight-deck director signaled us to taxi, and all fifty-six thousand pounds of our missile-laden jet slowly lurched forward, the nosewheel twisting and turning as Crocket expertly maneuvered it into final position. A green light flickered when the launch bar dropped in front of the catapult. With an eye on the yellow-shirted flight-deck director to my left, I turned to my right, making eye contact with another young Sailor on deck who was awaiting my orders. I gave two hand signals, indicating our gross weight, and he communicated back to me using an old flashboard you'd expect to find in an antiques shop in Virginia Beach. I gave him a thumbs-up.

The scrawny kid ran out from attaching the holdback fitting to the aft side of our nose gear. The sun slanted hard across his face, glinting on a patch of peach fuzz on his chin. *Missed a spot shaving,* I thought. *Hope he's better with our jet than a BIC.*

The tiny holdback fitting was the only thing keeping our plane on deck, and if it wasn't properly attached, I was dead, and so was Crocket, and the Navy would lose one of its best toys.

No time to worry. Once the director signaled to put the jet into tension, we waited to hear the distinct *click* and feel the whole aircraft squat down. The squat was crucial. Any significant delay meant someone had screwed up and we should immediately abort the launch.

All right, yeah, here we go. The plane settled perfectly into its crouch, and I shifted my eyes to my rearview mirrors, ensuring the jet blast deflector, or JBD, was in place, lest we blow all the people and gear behind us off the ship and into the ocean ten stories below.

"JBD's up," I told Crocket. The flight-deck director waved two fingers indicating we're clear to run 'em up.

After Crocket went through his control checks and I monitored our instruments and error codes, the two Sailors in white jerseys gave the flight-deck director a thumbs-up, hunched down below our wingtips, and braced for the roar of our Pratt and Whitney engines. Crocket lifted his hand to his helmet and dropped it in firm salute, the final signal on our part. The shooter, glancing to make sure the lane was clear, knelt on one knee and in a dramatic gesture thrust his arm down the catapult toward a clear blue sky and shimmering sea off the bow. Receiving this final signal, Peach Fuzz lowered his raised arms, ducked down, and pushed the red catapult button.

A one-second delay followed—the longest second of a naval aviator's life. I pressed my helmet hard against the seat back behind me so I didn't tweak my neck when we rocketed. Head firmly in place, I continued to monitor our progress. *Are our systems working? Oil pressure, fuel flow, error codes. Check, check, check . . . when in God's name is this waiting going to end?*

Exactly seventeen minutes after the *Bush* detected an Iranian plane entering our airspace, our F/A-18 surged forward, accelerating from zero to 140 knots in 306 feet. Air knocked from my chest and vision blurry from pressure, I kept my eyes locked on our airspeed, altitude, and pitch attitude indicator. The aircraft hit the end of the deck and launched—the sensation of plowing through concrete—and my body snapped forward in a brutal whiplash motion.

I gulped for air as Crocket took control of the aircraft, flying the plane like he'd stolen it. He manhandled the jet into a huge bat turn, rolling and pulling eighty degrees, pointing our aircraft toward heading 330. The jet shook, air shrieking, white vapor pouring off its wings. We doubled back around Mother, the ship disappearing in our rearview mirrors, and I tuned our sensors ahead for a rogue Iranian airplane somewhere in the empty sky. *We're coming for you . . .*

CHAPTER ONE

★

February 13, 2014, Pier 14, Naval Station Norfolk, VA

Close to midnight on the eve of Valentine's Day 2014, the minivan's headlights swept around the pier parking lot on Naval Station Norfolk as Mere's mom looked for an empty spot. Like drop-off day at camp, the parking lot bustled with families unloading loved ones. Young men and women piled out of cars, many unsure, some crying, a few even tipsy, like us. Contrary to the ship drivers, as aviators, we didn't have anything to contribute to the mission until we were well under way, so it didn't matter if we were present in mind *and* body when we stepped aboard. So long as we shuffled across the quarterdeck before the stroke of midnight, we'd have twelve to fifteen hours to sleep it off and get our minds in the game.

I looked out the van's window, my head still spinning from the bottles of wine at our goodbye dinner, one of the many last hurrahs over the past few months. Most of us had been living like there was an impending apocalypse—dining at favorite restaurants, going out every night, bidding lovers and leases farewell, not worrying about the calories or cash we'd consumed at alarming rates.

"Caroline." Mere looked back at me from the front passenger seat. "Didn't you earn a spot to fly on?"

I nodded. After a year and a half with the squadron, my seniority entitled me to a coveted spot to fly a jet onto the boat, which would have given me a few more days ashore.

Mere's mom looked in the rearview, chiming in, "So why are you walking on?"

"Didn't want to miss dinner with my girls," I said, but the truth was that walking on the boat was easier to plan. When you fly on, the Navy gives you a three-day departure window, but no firm embark date. In preparation for deployment, I'd already vacated my roommate Ashley's house and put all my belongings and car in storage. With no spouse or kids at home, there was nothing to keep me lingering on shore.

"Hang on." Mere drew a buzzing phone from her purse. "Hey!"

All night we'd been getting calls from friends and family, frantically saying last goodbyes. Aunts, grandparents, old boyfriends, long-lost neighbors who'd heard the little girl down the street was headed to war.

The minivan lurched to a stop, and we stepped out into the brisk night, watching the Sailors stream toward the mountain of steel looming alongside the pier. The USS *George H.W. Bush,* the Navy's newest Nimitz-class aircraft carrier—a super carrier—was over a thousand feet long and rose 134 feet from the waterline to the top of the flight tower that glowed orange in the sulfur lights of the pier.

We unloaded the bags in silence until Johanna, a helicopter pilot boarding the boat with us, stopped, discovering a hidden cooler of beer. "Hey, what's this?" she asked, her tone lightening the somber mood.

Mere's mom shrugged. "Heard it was customary to chug a

beer before walking aboard," she said. "Figured we can't go breaking tradition . . ."

We passed a few beers around and my mom chimed in, "What's that thing you do . . ." She threw her hand back, gesturing. "What's it called? Shotgunning?"

Johanna shook her head. "Pass me the car keys."

One by one, we cut holes in the bottom of the cans, lifted the ice-cold beers to our mouths, and popped the tops. An explosion of suds and laugher followed. I wiped my chin and looked at my mom, splattered with beer. As she lowered the dripping can from her mouth, I realized her smile had turned to sobs. This was more than saying goodbye and she knew it. She'd been so strong up to this point, watching me build toward this moment. Years of training and goodbyes had brought us to this, and after a week spent packing and cleaning my house as I scrambled to get my life in order, she crumbled before my eyes.

"Mom," I said in my most soothing voice. "It's gonna be okay."

I've never been physically affectionate, even with family and friends, and my time in the Navy certainly made me less so. Still, I knew then my mom needed me, so I stood there, hugging her tight. It was a big moment, but I couldn't cry. As though in an emotional dive-bomb, I let the excitement and fear and sadness propel me at full speed, resisting my survival instincts to just pull out. To go back to a life that was predictable and safe.

I couldn't help but look over Mom's shoulder to the boat and the adventure it promised. I could think of nothing else. The Navy, if anything, prepares you for departures.

CHAPTER TWO

★

June 2005; Annapolis, MD

Annapolis, Maryland, is a too-cute-to-be-real colonial town on the Severn River, which flows into the Chesapeake Bay. It is part capital city, part college town, part sailing mecca. The kind of place where officers in crisp Navy whites mingle with prepsters in pink seersucker pants in a setting that could be featured in *Travel + Leisure.* Gorgeous in June, but with the heat and humidity, the air practically boiled.

Induction Day at the United States Naval Academy (or I-Day, as it's called) commenced at exactly 0600, so it was predawn when my parents pulled the car through the heavy iron gates and my journey as midshipman began. My older brother, Craig, had just finished his sophomore year at the Academy and was deployed for summer training on the other side of the world, so my send-off party was small, and the morning rapidly moved ahead with little time to cling to the moment.

"I've got to line up now," I said cheerily to my parents.

"Well." Mom pulled down her sunglasses. She was far tougher before she'd sent her babies off to war. "Good thing I've said this kind of goodbye before," she said, remembering, of course, when she left Craig at the Academy two years prior. My father wasn't as stoic, his shirt wet first with sweat, then tears. Ill-practiced at hiding his emotions, he visibly shook until someone patted him roughly on shoulder.

We turned to see Vice Admiral Rodney Rempt, the three-star admiral and the superintendent of the Naval Academy, eyes smiling out from behind salt and pepper brows. "Don't worry," he said. "We'll take care of her."

Two big hugs and a couple of curt commands later, I crossed into the Academy wearing a coral-colored Polo and carrying a bright floral-print bag filled with the amenities we were instructed to bring and nothing more. The rest of my personal effects for the summer would be issued to me.

Along with a confused class of 1,200 other plebes, I spent the rest of I-Day passing through a labyrinth of tests, vaccinations, check-ins, and gear fittings, all at a brisk pace while the cadre screamed in our ears. As I lugged my hundred-pound gear bag up the stairs of Bancroft Hall, I felt my saggy, newly issued pants slipping off my waist. I didn't dare drop my bag for fear I couldn't pick it back up, and I was determined not to start out by asking for help.

"Johnson, what the heck!" the cadre behind me screamed as I felt a whoosh of cool air. "You're mooning your entire class!"

Around six p.m. that evening, I took the oath of office. Just twenty-four hours before, I was a somewhat privileged Colorado debutante with doting parents, enjoying all the freedoms of an American eighteen-year-old. Now, I was no longer even a civilian; I'd become property of the United States Navy. After the formalities of the day, we were lined up once again and ordered to march. Hustling through the massive doors of Bancroft with the rest of my classmates, I noticed a child off to my right, pulling her mother's sleeve. "Look, Mommy," she said. "A girl one."

Welcome to the United States Naval Academy.

★

The term *plebe* is derived from the ancient Roman word *plebeian*—a commoner, a member of the lower class. So naturally, at the Naval Academy, plebes are the bottom of the barrel, the lowest of the low on the totem pole. Commencing with I-Day, followed by six grueling weeks of Plebe Summer, all freshmen experience the unrelenting scrutiny of not only the cadre, but the resident officers, senior enlisted, and civilian staff at the Academy. Basically, everyone is waiting for you to mess up. And let me tell you, as a plebe, you will mess up. This intense scrutiny is part of an intricate system designed to make students fail. Here you have twelve hundred of the smartest, most athletic, highest-performing college freshmen—many of whom have never been bad or even average at something. The pressure cooker of Plebe Summer ensures that each midshipman will become intimately familiar with failure.

Two and a half weeks in, I no longer knew myself as Caroline, but as 093258—my alpha code for the next four years. The identity number was just one part of the conforming. To meet Navy regulations, my long blond hair had been lopped off just days before my arrival, but still I found myself reaching for it like a vestigial limb. I started to realize that in becoming a midshipman, the outgoing, fun-loving Caroline from before wasn't much different from my thousand-plus identically dressed, pungently smelling, stressed-out classmates. Everyone was focused on survival.

As the grueling summer ground on, we ran, drilled, and silently absorbed an endless barrage of ranting. Of course, we were restricted from leaving or making phone calls, but even within the Academy walls, who we spoke to and how we spoke to them was also limited. The isolation was so severe that when my grandmother died, I didn't even know it until Craig found me during Sunday chapel. Knowing it was one of the few places a cadre could not snipe you, he summoned me out of my pew.

"This way," he said, leading me down underneath the chapel.

"What's going on?" I whispered.

"Wait here," he said and left me alone with an elaborate gilded coffin which supposedly contains the remains of John Paul Jones, the father of the US Navy and presumed pirate.

A few moments later, Craig returned with a Navy chaplain. "Caroline, I've got some bad news," he said.

It was the first time I'd been addressed by my first name in four weeks. I, of course, broke down and cried when I heard the news. We weren't overly religious, so at first I didn't understand why Craig had invited the chaplain into this very private moment. But as the chaplain began to comfort me, I realized that in his quiet and stoic way, Craig knew I would need someone to talk to, someone more experienced in life and death and perhaps more delicate than him.

In the wake of my grandmother's death, the rest of Plebe Summer loomed before me like a hundred-foot wall, so tall and daunting one could never imagine making it to the other side. But by focusing on climbing and not the barrier, the weeks marched forward in quick succession, and before we knew it, our company of forty strangers had transformed into a cohesive, efficient team of thirty-six. We could square our corners, recite any line from *Reef Points,* and run through an obstacle course with more dexterity than an American Gladiator.

I qualified as a sharpshooter on the M-16 rifle team and could hand-strip and wax a floor to make it shine. But perhaps more difficult than all of this was maintaining my femininity. The Navy seeks to both toughen and homogenize. And some people at the Academy went to great lengths to homogenize me. The result was the opposite. I fought to cling to the things that made me Caroline. Small things, like the color of nail polish, the cut of my hair—even if short, the unmistakable glow of shoes made from Italian leather. These small things mattered to me. They make me happy.

Feeling like a woman makes me happy. And I had feminine examples at the Academy to look up to like Ashley, a beautiful and bubbly classmate of mine who was in my brother's company. Driven like the rest of us, Ashley had a magnetic smile, grace, and femininity that was uncommon at the USNA. Had it not been for her father being a senior officer in the Navy, I don't think Ashley would have been the kind of girl to wind up there. But she thrived and quickly became one of my first girlfriends at the Academy, a girl who both encouraged and inspired me to hang on to those small things that made me who I am.

But Ashley was the exception. Most people at the Academy, including midshipmen and teachers, want to force homogenization. For some, my determination to maintain my identity—even in those very small ways—became a focal point of their lives.

More times than I could count, one such person would have me stand at attention outside her dorm room in Bancroft Hall (the largest single dormitory in the world, housing the entire four-thousand-member brigade) and shout, "Midshipmen Uniform Regulations—*MIDREGS* Article 1106.1-1. *Conspicuous*: items that are obvious to the eye, attracting attention, striking, bright in color should blend with, not stand out from, a professional appearance in uniform—"

"Louder, Johnson!" Bonnie would snipe, her mouth inches from my face. Bonnie was a firstie—the civilian equivalent of a senior—who had, it seemed, decided to make me her pet project. After only a few weeks, I knew the drill. She'd make me read the Academy rulebook, *MIDREGS,* until I knew every word right down to the punctuation points. But in forcing me to learn the rules, she'd also taught me how to manipulate them. For instance, Bubble Bath Pink, I discovered, was a color of nail polish that abides by *MIDREGS* 5102.4.b. And if I wanted to wear Italian pumps instead of the Navy's old-lady shoes, I could have the heels taken down by Navy cobblers to two and five-eighths inches. I also had a Navy tailor tuck my baggy, black uniform down to the nanometer, so it fit per regs but without the "bag of leaves bunched around your ass" effect the Navy seems to require of its female population.

"I said louder, Johnson! We're not done yet!"

"Ma'am, yes ma'am! 'What is conspicuous on one person may not be noticeable on another. If attention is naturally drawn to or distracted from the professional appearance, it is conspicuous . . .'"

This kind of harsh refinement went on until the culmination of Plebe Year—a capstone event known as Sea Trials, usually conducted mid-May. The day begins at two a.m. and continues for fourteen hours. The physical and mental challenges include: simulated emergency resupply, short offense, a Spartan relay, a combat fitness test, obstacle courses, simulations of urban terrain combat, pipe patching, fire-hose handling. There are also water events: tests on the water, in the water, underwater, aboard boats, in the mud. There is a hill assault. A two-mile run. Bridge defense and demolition, paintball and jousting. A rucksack run and simulated evacuations. Basically, for fourteen hours straight, you're crawling through dirt, weaving through obstacle courses, doing a lot of physical training—or PT, as the military calls it. It's not unlike what Navy SEALs might endure during their BUD/S class, albeit on a lesser scale. The reward for completing Sea Trials is a well-deserved dinner—an Academy picnic with family, classmates, Sea Trials detailers, and senior leadership.

At the end of the day, I was wearing camo fatigues, covered in dirt, absolutely spent. No makeup and my hair looked like it belonged to a dog after a mud bath. My family was there—my mom, dad, even Craig. After hellos and hugs, I gave in to my ravenous hunger and was inhaling a barbecue sandwich when Vice Admiral Rempt once again approached our family.

We rose to our feet, but he motioned us to sit.

"Sir." He reached over and shook my dad's hand. "I told you we'd take care of her, and actually, she's done a pretty good job of taking care of herself . . . she's one tough girl."

I blushed at the compliment. "Thank you," I said, covering my mouth in case there was any barbecue in my teeth.

"Excuse me?" the admiral's wife chimed in, a look of shock on her face. "Is that a fresh manicure?" She pointed to my fingernails, caked in mud, but the polish unmistakably pink underneath the grit.

I really started to blush. "Yes, ma'am. Painted them yesterday."

She turned to her husband. "Only twenty-two of twenty-eight of their company completed the course today, and Caroline did it without breaking a nail!"

★

May 22, 2009; Navy Marine Corps Memorial Stadium, Annapolis, MD

Sweat dripped down my back and soaked my shirt, tucked into my white polyester dress skirt. Since we were forbidden to wear sunglasses onto the graduation field, I squinted into the blinding sun that caught the sea of crisp white uniforms. We, a class of 1,057 firsties, stood at attention on the sweltering field in the Navy Marine Corps Memorial Stadium. You could almost hear a collective sigh of relief. Four years of the grueling and transformational officer factory known as the United States Naval Academy and we'd come out on the other side.

The stadium was packed, every seat filled with family, friends, and honored guests who awaited President Obama's arrival. It was only ten o'clock in the morning, but the air was more than hot, it was drinkable, and smelled of booze and sweat from our weeks of celebrating. Many families had saved for four years just for graduation, renting one of the beautiful, historic homes in downtown Annapolis for the week. And as it did every year, the quaint coastal town overflowed with rejoicing firsties and their families, tourists, and Sailor-preppy locals in pink pants and whale belts.

Like everyone alongside me, I'd started my journey as a plebeian, but there I stood, a graduate of one of the most prestigious and challenging universities and a commissioned officer in the United States Navy and Marine Corps. In that moment, sunblock and makeup mingling with the sweat on my forehead and streaming into my eyes, I felt like I'd climbed Everest. Step by painful step, I'd started up an uncertain path, and now I found myself looking out from the summit.

The feeling was exhilarating but fleeting. The Academy was just the start. I knew many far more difficult and dangerous mountains were out there. We all tried to downplay it, but my classmates and I also knew the specter of war was a real possibility for all of us, some of us in planes, some on boats, and some with boots on the ground—or, more accurately, the sand.

After President Obama spoke, we filed on stage one by one and shook his hand. I made my way up the ramp, grinning and gracefully balancing in the two-and-five-eighths-inch Italian heels I'd custom ordered for the occasion. The president had been uncharacteristically formal and somber in this, his first service academy graduation. As instructed, I reached for his hand, but before

I could utter the mandatory phrase, my excitement got the best of me. *What do I have to lose?* I thought. *The Academy's not going to kick me out now.*

"Sir," I said, catching the president's eye. I pointed to the stands where my parents, Grammy, aunts and uncles, and Craig sat watching. "Will you wave to my family?"

His tight demeanor uncoiled and he laughed in his familiar way, turning to wave at the twenty-five or so now roaring Johnsons.

"Congratulations!" He beamed.

"Thank you, Mister President, sir!"

I exited the stage and Ali clobbered me. "Caroline! We made it!" My best friend smiled.

"Can't believe you got him to wave to your parents—you're ridiculous! Now he's fist-bumping people."

We braced for the cadence of the three hip-hip hoorays led by our class president. On the last hooray, we exploded toward the sky as one, each throwing our midshipmen hats—our covers—as high as we could as the Blue Angels screamed by in their Delta formation, so low it looked like our covers might hit them. I stared up, mesmerized by the precision and power of the F/A-18s, their pattern as consistent as floor tiles. As the perfect white contrails faded into the atmosphere, I pumped my fist in their direction. "Hell yeah!"

★

After graduating from the Naval Academy, many new graduates rushed from the structure and stability provided by Bancroft Hall, aka Mother B, to another potentially mothering institution—marriage.

So every year, the Academy chapel became a veritable nuptial turnstile, booked from two hours post commissioning ceremony for a solid two weeks. With the clockwork precision that marks life at the Academy, ceremonies are conducted fifteen minutes apiece, with fifteen minutes in between to usher one crowd out and another in.

I personally struggled to understand how anyone who'd yet to experience the strange joy of making themselves an omelette, or even waking up alone in their own apartment, would yearn for the added dependency of a spouse as they transitioned into the unknown. I watched as the last days of USNA turned into a game of musical chairs after the music phased out. For some of my classmates, it didn't really matter who they were dating at the time, when graduation rolled around, they scrambled to the altar to settle down.

But they weren't all that way. The last event of my graduation week was the wedding of one of my best friends, who was very much in love. The ceremony was quick and sweet, and feeling a bit partied out from the week of festivities, I planned on dancing a little, then sneaking out. Tired as I was, the former Colorado debutante in me could never pass up the chance to get dressed up and whirl around on a shiny floor. The question was, *who to dance with?*

For four years, I'd been surrounded by hundreds—actually, thousands—of eligible men. Handsome and charming, smart and fit, ambitious and polite. But looking around the reception, the problem was glaringly apparent. I'd grown so close to them that they felt like brothers. Sweaty, drunken brothers at that. Like the proverb says, familiarity breeds contempt, and that was just fine with me. I walked through the doors of the ballroom with my usual guard up and my hair down.

I was dancing in a circle of friends when over by the bar, a man's torso caught my eye. I don't make a habit of checking out men, but the oversized chest was impossible to miss. A perfect upper body attached to stick-thin legs. *Damn,* I thought, inching my way over for a closer look. *It's Channing Tatum's twin.*

He was wearing the dark blue jacket of a recently commissioned Marine, and the unapproachable, thousand-yard stare of a combat veteran.

The stare was what placed him for me. He was the bride's cousin, who I'd met briefly at a house party a few years back.

"What's your cousin's problem?" I asked my friend as we left the party later that night. "Never seen anyone's face so deadpan."

"Iraq," she huffed. "He enlisted while he was at VMI. He was in the heat of the firefight in Fallujah and many of his close friends didn't come home."

There at the wedding, her words circled back to me as I watched him.

I felt for him, but I couldn't help but notice how attractive he was. *Hot,* I thought. *But better leave him alone.* Still, I couldn't stop staring at his chiseled upper body and skinny legs. *Built like a Minotaur.* I smiled to myself.

A few of my classmates, also freshly minted officers, clustered at the bar, and Minotaur stood just beyond them, a beer in hand and detached look on his face. *Screw it.* I sauntered over, trying to keep cool, my mind spinning with what to say. He gripped a long-necked beer and pretended not to notice me heading closer.

"Having fun yet?" I asked, immediately regretting it, realizing he couldn't hear my high-pitched voice above the crowd. I nervously tucked my hair behind my ear and tried again, louder this time. "Too early in the night for a Bud Light."

His eyes shifted to me, squinting, but not in a friendly way.

"Thought Marines preferred hard stuff to start the night." I carried on. "Not up for it?"

I saw the chink in his armor as he let out a short laugh. "I've been here, surrounded by dudes all night . . . and I finally meet someone from the Academy with a pair of balls." He put his beer down and held out a hand. "It's Caroline, right?"

"Two bourbons—neat." Minotaur signaled the bartender, then turned to me, one muscular arm resting on the bar. "So which one of these dudes staring at me right now is your boyfriend?"

I shook my head. "None of them."

"Gotta be exes then."

"Nope." We clinked glasses. I downed my bourbon and signaled for a second round, catching another grin.

"Then what's that dude's problem." He nodded at my friend Ryan, a newly selected SEAL, leaning against the wall, blatantly glaring. "Looks like he wants to bash my head in."

It was true. Whenever a new guy approached, my brothers from school instinctively fell into a combo protective/competitive mode.

"He probably does," I said, sipping my second bourbon slowly. "Better not turn out to be a creep."

"We'll see." Minotaur shrugged, knocking back his drink. "Like to dance?" He offered his hand, knowing I'd take it.

Twirling around the polished wooden surface, the room fell away. We may have been sharing the dance floor with the bride and the groom and dozens of others, but I'm not sure either of us were completely aware of those around us. We danced, silently absorbed in each other. Eventually he introduced me to his sisters and parents, but I barely noticed because right afterward he leaned in and asked, "Wanna get out of here and head somewhere downtown?"

"But your family is here."

"As much as I love the Cupid Shuffle," he said, "the stares from your Academy buds are burning holes in my back."

Strolling the downtown streets alone, in the cool of the night, I could see the Minotaur's shoulders ease back. He relaxed and with every step, opened up a little more. We talked and laughed, dropping into a few bars. He asked me if I liked cigars (which I do), and we shared one on the dock before he walked me back to the house my parents had rented.

"Come inside for one last drink?"

He gave me an uncertain look. "Your parents are in there?"

"They're asleep. I have bourbon."

He laughed, and stepped in. I poured glasses and we sat together on the couch, somewhat awkward now.

"Yeah, I have a mundane desk job until I go to flight school," I told him.

He nodded. "I hear ya. I'm doing land-nav training at the Marine Corps Basic School, about an hour and a half from here."

It was almost dawn and we both found ourselves yawning, adrenaline fatigue of a long week.

"Better go," he said finally, pulling me to him into a tight embrace. "Hope to see ya again, buddy." He grinned and left.

I wasn't going to ask if he would call or if we'd see each other again. I knew no matter what, we'd likely meet in Florida, where he'd be training to become a Marine helicopter pilot, and I'd be pursuing my career as a naval flight officer.

After he left, I headed to bed and tried to divert my thoughts from this new boy, the first guy I'd met and liked—really liked—in a long time, and I had the strange feeling he felt the same way about me.

But so what? We're in the military. Anything can happen. It was only a matter of minutes before I crashed, body and soul, into a deep sleep.

CHAPTER THREE

★

February 14, 2014; Embarked USS *George H.W. Bush*, Norfolk, VA

While our adversaries can do little to stop one of America's largest nuclear-powered aircraft carriers, a microscopic, worm-like bacteria in the James River can. Unbeknownst to the aviators aboard the *Bush*, while we slept off our reverie, the worms entered the intake valves, expanded, and clogged up the reactor. As the nuclear engineers aboard started the reactors to depart on deployment, they uncovered the issue and spent the early-morning hours cleaning out slimy, stringy parasites by hand so we could depart at the next high tide. Thus, we delayed until the afternoon and set sail on Valentine's Day, 2014.

Once we were free of our pier tie-downs, our deployment technically began, but before Mother could start her journey across the Atlantic, a few items of business remained. First off, the carrier cannot leave the harbor loaded with aircraft, so we pulled out with an empty flight deck. Once we got out to sea, the entire air wing—helicopters, E-2s, C-2s, and all the jets—would have to land onboard the ship and all the aviators had to carrier-qualify, or CQ, before we could leave the coast.

We all had previously carrier-qualified in our training, and for some, in previous deployments, but landing on the boat is so dangerous and such a perishable skill that every single time you go back out to sea, you have to requalify. There is a three-page eye chart that dictates how many landings aviators must acquire if they've gone more than seven or fourteen days since their last boat landing. Since it had been more than sixty days for us, each aviator had to do four day landings and two night landings to requalify before we could depart US waters. This may sound simple, but it's a fairly high hurdle since it required more than 145 aviators to cycle through fifty-three fixed-wing aircraft. With so little space on the flight deck, during qualifications, planes would fly back to Virginia Beach to gas up and wait for sunset to night qualify on the boat. Each time aviators went ashore, they loaded up on junk food—Taco Bell, McDonald's, Chick-fil-A—all offerings for their hungry counterparts aboard the ship. In no time, I'd consumed so much junk food that I felt like I'd been given a blood transfusion of pure hydrogenated fat and preservatives. Whenever a plane returned from the beach, I looked to the sky and prayed for salad.

In addition to the carrier qualifications, the strike group experienced an airborne change of command, meaning the admiral who commanded our air wing and twelve warships during our most recent training would turn over the reins (in midair) to a new admiral for deployment. A change of command at sea is rare, but rarer still is one conducted between two admirals who flew the jets themselves. A special, but somewhat melancholy event, the departing admiral, who we'll call Lung, was beloved. He'd trained and led us over the past year, fine-tuning our team so that we were ready to face any adversary. Lung was an incredible leader, and a downright good guy, with a well-known affinity for late-night tequila shots at the Officers' Club. All of us felt a little heartbroken when the call came for him to move on to his next tour. We hadn't yet met the new admiral (who I'll call Bullet, because he was small but his personality packed a punch), but we'd heard good things.

Instead of the standard pomp and circumstance of a shore-based change of command ceremony, where salutes and handshakes usher out the old and welcome in the new, in an airborne change of command, the admirals symbolically exchanged power through the maneuvering of jets. Well, not just any jets—F/A-18 Super Hornets.

Approaching the ship, the men flew their Super Hornets in tight formation, low and fast, like they were ingressing a target area in combat, Lung leading and Bullet following close behind. Nearing the stern, Lung passed the reins via hand signal and Bullet pulled ahead in a clear lead change. As they tore past the flight deck, Lung ripped back on his stick, sending his jet's nose straight up. Bullet continued the course straight ahead by himself, and Lung shot into the atmosphere so fast and so high that he disappeared from sight. Flying alone, Bullet circled back to land on the carrier. As tradition dictates, once the torch was passed, out of respect for the new leader, Lung was not spoken of again.

On the flight deck, we cheered and welcomed Bullet as our new Strike Group commander. If a command change of that level had happened back in Virginia Beach, it would have drawn crowds, hundreds if not thousands. Spouses and VIPs and all their followers would have gathered for the spectacle, but on the boat, the airborne change of command was reserved just for those who had access to the flight deck. Intimate and exclusive, it felt a bit like we accidentally were invited to a celebrity wedding. Meanwhile, the rest of the four thousand people onboard, scurrying somewhere below like mice in a maze of steel, only learned of the shift in command by the tolling bells and an announcement over the 1MC loudspeaker when Bullet's jet landed.

With a new leader in charge and the entire air wing carrier-qual'ed, we finally set sail for the open sea.

CHAPTER FOUR

★

Summer 2009; Annapolis, MD; Newport, RI

The Naval Academy is tuition-free, paid for by American tax dollars, so thank you. As a graduate of the Academy, I owed the Navy (and taxpayers) a minimum of five years of my life, and because I chose aviation, a few years more than that.

To repay the debt, you worked, and like my dad on weekends, the Navy had no trouble finding random work for you to do. Some of it's interesting, but a lot can be the mind-numbing, bean-counting type I first experienced in Annapolis right after graduation. Inputting USNA student banking information into the Navy's antiquated computer system might not have felt like what I signed up for, but it was a means to an end, and for me, that end was flight school.

In aviation, the first real test for future aviators is Introduction to Flight Screening, or IFS, a short program where ensigns learn to fly small civilian Cessnas while the Navy determines if you have the aptitude and drive required for the vastly more complicated and expensive military training aircraft.

Once again, it was steaming hot in Annapolis, but I didn't care because I was finally in a cockpit—my feet steering the pedals, yoke in my left hand, throttle in my right. Instantly, I was hooked and took to class like a junkie looking for her next fix. At the time, the IFS program paid to put students through twenty-five hours of training, just shy of the thirty-five hours required for a private pilot's license, but with the encouragement of my instructor, I paid out of pocket to finish the final ten hours I needed to take my private pilot check ride. Passing the test thirty days after starting, I officially became a private pilot.

Having my first taste of flying, I dreaded returning to my data-entry desk job in a baking-hot Academy cubicle. Luckily, the Navy had other plans for me and sent me to teach mathematics at the United States Naval Academy Preparatory School in Newport, Rhode Island, a seaside resort town dotted with summer cottages built by the industrial titans of the late nineteenth century. My students were mostly athletes, and at risk of sounding cliché, they were, in fact, mostly football players. In later years, when I would see my former students playing in the famous Army–Navy game, I felt proud, having helped the Navy kick the Army's ass in some small way.

While I discovered a love for teaching, I was ready for the next phase. The European beech trees in Newport hit their peak, a blazing riot of orange and red, and as if on cue with the seasons, I got my orders to leave. The Navy instructed me to migrate south and report to flight school in Pensacola, Florida.

Before arriving, I had a few logistics to sort out, chiefly, my home for the next two years as a flight student. There were three housing options available to me: living on base (an immediate no), pitching in with a couple of roommates and renting a place off base (maybe, but I wanted my own space), or buying/renting a house to fix up and call my own (too expensive). A fourth, albeit unconventional, option presented itself while I was visiting my grandparents on Torch Lake, Michigan. There, parked in the weeds, was Grammy's 1969 Avion camper trailer.

Every year of my childhood, as soon as school was out, my family loaded up the minivan, complete with kids, dogs, and an occasional hamster, and headed to Torch Lake for the summer. We spent our days waterskiing, canoeing, wakeboarding, sailing, roaming the woods, playing board games, and doing chores for Grammy and Grampy. While my grandparents were great sports about the gang of kids taking over their home, after a day of providing chocolate chip cookies and using their expertise as a doctor and nurse to patch up medical emergencies, they would quietly sneak out to their "tin can" to escape the confusion. They would turn on the Avion's antiquated air-conditioning unit to drown out the noise and hide until the next morning. After Grampy passed, the Avion was mothballed and locked up, and became a repository of dead chipmunks and mold. We were all having breakfast on the porch when the idea came to me.

"Grammy," I asked sheepishly. "Think I could take the Avion to Pensacola? I'd fix it up, of course. See, I'm trying to find a place to live alone and don't have many options . . ."

"The tin can?" Grammy stopped to think. "Well, sure, if you want. Long as you promise to take care of her. Right now she's just gathering dust, sitting as it is."

"Mother," my dad said condescendingly. "Caroline can't move that thing, much less live in it. It's a liability. More burden than blessing."

Grammy winked at me, knowing a beautiful relationship had just begun.

★

My dad did have a point. The tin can lacked many comforts of your typical apartment. Only 220 square feet, it looked like something you might see in a 1960s nuclear test video. But after four years living in a cramped USNA dorm room, surrounded by enough people to warrant a municipality, the trailer and the privacy it afforded was a luxury. The surfaces of the Avion might have been moldy and rust covered, but its bones were solid. As I studied her beautiful lines, I could see her as my home, remodeled inside and out and parked next to a sandy white beach.

The problem was finding a place in Florida where I could restore her. I started calling trailer parks to inquire about a slip, fingers crossed that one of them met a few criteria. I wasn't asking much, just something close to the base, perhaps with a waterfront view, and hopefully free of meth labs. To my surprise, I found plenty of nice options, but all were booked up until spring and each one wanted to see my rig in person before making a decision.

"I know it's policy, but she's a 1969," I said to the woman on the phone. "A classic. She's going to be beautiful. Can't you just take my word for it?" I never would have guessed that a trailer park would be pickier than me.

"Need to see the rig," the woman on the phone said curtly.

"Then what do you suggest? A junkyard?"

"That'd be a start," she said, and hung up.

This was going to be harder than I thought.

CHAPTER FIVE

★

February 15, 2014; Embarked USS *George H.W. Bush*,
Warning Area 72, Atlantic Ocean

As I had learned at the Academy, room selection and roommate pairings are important and cutthroat in the Navy. Lucky for the ladies of the Sharktank, we'd scoped out our accommodations on the multiple under ways before deployment. We knew exactly where onboard the ship we wanted to be, and so we put our request in early. It helped that the Carrier Air Group commander—the man in charge of all twelve hundred people attached to the nine squadrons aboard—had raised daughters and knew that if his girls weren't happy, no one was happy. So needless to say, we got our stateroom of choice.

As officers onboard, we were lucky. Even though six women living in 350 square feet seemed cramped, our enlisted were housed in massive dormitories called berthings. They shared one room with two hundred others, the beds were stacked three deep. When you looked at it that way, five roommates wasn't so bad. On the boat, the only people with their own rooms were our squadron commanding officers, XOs, and a high-ranking foreign aviator I would come to know well.

As senior as these men were, even they had to share bathrooms. The admiral, the ship's commanding officer, the CAG, and a few other officers had their own staterooms and bathrooms, but being in charge of 7,500, 4,000, and 1,200 people, respectively, they'd earned it. There's a saying in the military: "rank has its privileges," and there's no place where those words ring truer than on an aircraft carrier.

Our stateroom, located two decks below the flight deck, was a definite luxury. Most aviators, including top brass, had rooms directly below the floating runway. If you've never heard a jet landing or launching off a flight deck at night, imagine standing next to a brick wall as a car, going 150 miles per hour, crashes into it. If that's not earsplitting enough, think of being trapped in a dumpster with the entire Boston Red Sox team beating on it with aluminum bats. I'm not exaggerating; living and working directly below the flight deck is that loud. Ricocheting and intensifying through the metal, the sound is deafening, so the Sharktank, tucked below the racket, was prime real estate.

Just like we'd asked for the room early, we'd also meticulously planned its aesthetic. In the two months we lived on the boat during workups, we spent our evenings watching HGTV and planning like tiny house designers, writing shopping lists and drawing sketches for the space. After months of Amazon deliveries, packing, and hauling our goods onboard, our deployment had finally begun, and we were ready to put it all together. Given that we had to cram everything we needed into a room that was not much bigger than a walk-in restaurant cooler, we had to get creative to make all the gear fit. This involved the six of us channeling our inner Bob Vilas and using our dads' carpentry skills to install custom-built laundry shelves and storage platforms. We weren't halfway across the Atlantic before we were dialed in and had everything in its place.

The Sharktank was divided into two parts by a bank of

closets that ran down the middle of the room. On the side closest to the outer wall of the ship was a narrow walkway with six racks (Navy term for bunks) lining the sides. The racks were stacked two high and adorned with curtains made of designer fabric that one of the Sharktank moms had sewn for us. On my bed, I had three memory foam pads over the standard, government-issued mattress, and my sheets rotated between a Russian ballerina print that I bought on a trip to St. Petersburg and a colorful, geometric pattern. Even within the Sharktank, my rack was my retreat. The only space on the ship that was truly mine. A comfortable, happy oasis where I could be alone, and the introvert in me could escape from the constant stream of people.

The other side of our room was our living space. Each of the six of us had a rigid, metal chair and desk that dropped down from our storage locker, like a Murphy bed. The desk was just large enough to fit my Macbook Air and a small cup of coffee, but nothing else. There were also two aluminum sinks and medicine cabinets for our toiletries, the shelves of which were lined with professional-grade OPI shellac nail polish in every color. A man I'll call Lumberjack, in his short tenure as boyfriend, had helped me set up a movie projector, a ten-foot drop-down movie screen, and he'd even installed surround sound. The walls, of course, were decorated with colorful decals and posters that transformed the space from boring battleship gray to a cheerful, girly space. I kept a hydroponic herb garden next to my desk and the boot dryers to air out our leather boots. We all had throw rugs under our chairs—mine was a whimsical Trina Turk—and one of our best purchases was a cordless vacuum for the endless strands of hair that tended to build up on the carpet.

Like a Native American with a buffalo, when it came to space, we wasted nothing. Six yoga mats lined the wall, sets of dumbbells were stacked in corners, and a rack was suspended midair with six matching laundry baskets secured on top. We even had our own coffee and tea bar with loose-leaf tea, a kettle, and a Keurig coffee maker with French vanilla creamer and every kind of sweetener. A few of our squadmates' wives even went to the trouble to send packages every month, replenishing our lip balm and magazines. Hard to imagine how they took time to ensure we had the girly comforts while they were juggling broken dishwashers, kids, jobs, all on a meager military salary. With all of the little luxuries of home, we were somewhat able to escape the industrial, stark environment of the rest of the ship.

Though the comforts of the Sharktank were great, my five roommates were the best part. Our group of six was comprised of three aviators who flew the E-2 and three who flew F/A-18s. The E-2 is not a jet, but a turboprop plane used for early-warning detection. It's a support aircraft, equipped with a huge radar that serves the very important role of coordinating the fighters in both Air-Air and Air-Ground employment. Basically, they were the quarterbacks of the air wing, but they were also saving our butts, should we get ourselves in precarious situations. The E-2 squadron on the *Bush* was known as the Bear Aces and my roommates were nicknamed Blonde Bear, Little Bear, and Tall Bear. Having the Bears in our room definitely served as a counterpoint to the more anal-retentive Jet Girls. Compared to us, the Bears were chill and easygoing, like longboard surfers of Tailhook Aviation and jet aviators were the shortboarders.

The Jet Girls earned the nickname "the Triple As" for our extreme type-A personalities but also for the acronym for Anti-Aircraft-Artillery fire—the deadly weaponry that we'd encounter in combat. The three of us were the initial members of the wolf pack: Carolyn, a single-seat F/A-18 pilot and the first and only female pilot in the history of her squadron since its founding in 1935; Taylor, another F/A-18 WSO, and me. Taylor not only was a WSO like me but we were also very similar. Both of us were good at being buttoned up and devoid of most emotion when we needed, but when alone we could laugh together, cry together, and best of all, be authentic in the widely competitive microcosm in which the Sharktank existed. Carolyn, Taylor, and I were as tightly strung as a perfectly tuned guitar, and since we couldn't control much on the boat, often our OCD tendencies manifested in cleaning.

Not surprisingly, the three Bears, strolling in and tossing their bags and papers on the floor, weren't on our same level of organization or cleanliness. Not only did they make the messes, they weren't likely to pitch in with the normal chores it took to keep the Sharktank up to standards. Let's just say if Goldilocks stumbled into the house of these Bears, she wouldn't have had a clean dish for her porridge.

After getting wound up over their sloppiness a few times early on, we realized we'd all have to adapt if we were going to survive the nine months together. The Jet Girls would need to chill out, and the Bears would have to neaten things up. Both happened rather quickly, and the Sharktank remained an escape, a little princess palace inside what could have been a giant, floating penitentiary.

Our stateroom quickly became legendary. From the admiral, to the ship's captain, to the lowest enlisted, everyone had heard of the epically appointed Sharktank, and like a far-off land in the days of yore, rumors spread about the exotic things that lay inside it. Since there were only two staterooms for all female aviators, everyone knew our whereabouts, and this, while fine at first, grew creepy the more often curious Sailors came to see it. Enlisted and officers alike would find excuses for passing by our stateroom.

"Heard you guys have a vacuum. We spilled a bag of coffee on our carpet, can I stop by to borrow it?"

Even though the enlisted weren't supposed to transit Officer Country (the restricted area in which we lived), often young boys passed slowly by our door, their necks stretching like Inspector Gadget. Or some of them on tiptoe by the exhaust vents, sniffing the lavender shampoo, fabric softeners, and French facial creams that poured from our room into the hallway. We wrote off the lingering, figuring the novelty would fade, not realizing it would become troublesome for the Sharktank.

CHAPTER SIX

★

November 1, 2009; Naval Air Station Pensacola, Pensacola, FL

Our first phase of flight training, Aviation Preflight Indoctrination, or API, didn't start right away, so the Navy found ways to help us pass the time. Rigorous physical testing, medical evaluations, basic leadership classes, random odd jobs—the three hundred plus USNA grads and I stayed occupied. I knew the interim work might involve menial tasks, physical labor even, but there was one thing I hadn't anticipated—mice turds.

Thankfully, I wasn't in it alone. Ali, my best friend from the Academy, and I were tasked with running the gedunk, the naval snack shack that traced back to days of black powder cannons. We might have seemed an unlikely team—Ali, a feisty five-foot-one-inch brunette from Southie and me, a tall blonde from suburban Colorado Springs—but the one thing we had in common is that we didn't put up with shit, literally.

The manager of the gedunk who preceded us was the typical flight school man-child who would rewear dirty sweat socks rather than taking them to the Laundromat, so it was no surprise that when we inherited the snack shop, it was a couple of dead cockroaches away from condemned. Undoubtedly, his predecessor had been just as slovenly, going back, I surmised, to the days when tall ships first spotted Florida.

Immediately, Ali and I closed the gedunk to the public and, donning the closest things we could find to hazmat suits, got to work. Not only was most of the food months—or years—past expiration, much of it had been sampled by resident vermin. We set up giant fifty-gallon trash cans and with Beyoncé boss-lady music blasting on the radio, we jump-shot our way through the old inventory. Once the space had been cleared, we got down to hand-stripping, waxing, and when it came time to scrub the floors, we thought about calling in backup. Typically in the Navy, officers can call on enlisted Sailors to clean and perform routine maintenance on the facilities.

"But I don't really see the point," Ali said. "If we ask for help, it'll be the same type of slobs who've let this place go for the last twenty years. They'll just get in our way."

"So we do it on our own," I called from the inside of the mold-encrusted fridge. "The right way. We had four years of practice at the Naval Academy."

"Don't I know that," Ali said.

Ali and I might have had $250,000 USNA educations under our belts, but still we weren't above a little manual labor. We knew even then that *doing* is how the Navy turns junior officers—those in the ranks of ensign to lieutenant (zero to ten years of commissioned experience)—into jacks of all trades who are expected to take care of all tasks needed to run a squadron.

When the gedunk reopened a few days later, not only did all the junk-food addicts get their fix of every sweet and savory snack-food science experiment from Fritos to Fudgsicles, but those of us who wanted something a little fresher had celery and hummus, Perrier, and even fruit.

Our work paid off, and the profits benefited various programs and charitable events hosted by flight school. Sure, we weren't flying fighter jets yet, but we were happy to finally be a part of the naval aviation community. Keeping everyone fed between workouts and classes served the greater mission of flight school, and while some of our classmates were pouting that they weren't yet dogfighting, Ali and I took it as a chance to shine. It wasn't until later that we would find out that women officers in aviation are often assigned duties considered beneath our male counterparts, like coffee mess officer, as I would learn firsthand.

★

The gedunk was just the start of my HGTV-worthy renovations that summer. Before API started, I'd shipped the Avion from Michigan to a fenced-in Pensacola junkyard, but I needed a space that I could access at odd hours and begin her transformation. The question was how to move the trailer.

Before I'd ever realized the dream of a home on wheels, I'd gotten myself a sleek Infiniti sedan. I loved it, but what I needed then was a truck, or better yet, a friend with a truck. I was leaving the gym on base, thinking about who to call when a brand new Toyota Tacoma pulled into the parking lot and stopped a few yards from me. *Well, there it is.* I almost laughed. Jet gray with tinted-out windows and a shiny towing package off the back. *Perfect.*

I stepped over and tried to look through the dark windows, but the door opened before I could see inside. "Hey, buddy."

Immediately I recognized the voice behind the glass and tried to hide my smile as the Minotaur stepped down from the cab, ever so sexy in his perfectly tailored Marine Corps tan and green uniform.

"Like it?" he squinted over his sunglasses. "Picked it up this week."

Fast cars and motorcycles are not uncommon for newly

minted officers. Many of my friends whipped around Pensacola Beach in Porsches, Audis, and big-engine muscle cars. Most of us hadn't paid for school, we didn't have student loans, and we were finally making officer's pay (modest compared to bank analysts or hotshots at startups), but it was enough to have a little fun. This splurging on vehicles might have gone a little deeper than that. Our lives were going to be spent in transition, moving again and again over the next ten years. Rather than buy what would pin us down—couches, TVs, houses, and so on—we opted for things that kept us transient: fast cars, trucks, and motorcycles. Some of us took this too far.

"Didn't I see on Facebook that you bought a hot little Subaru WRX?"

"Yeah. Didn't like it." He shrugged. "Too small. A little girly."

"What'd you do with it?"

"Eh, they wouldn't take it back at the dealership, so I just traded it in."

"After a month? Bet they killed you on depreciation."

"Whatever. This is much more me." He kicked one of the monster tires and puffed his bulging chest. "I can even load my Harley in it." He winked. "Yep, picked up one of those, too. And you'll love the ride."

In the previous months, I'd seen plenty of my peers blow through the cash they'd saved during school. Finance classes weren't part of the Academy curriculum, so when we graduated, ensigns often overspent and overborrowed, some winding up in major debt. Car dealers love to see newly commissioned officers stroll into their sleek showrooms because they know the officers have steady paychecks, and if they default, their commanding officer will ensure the dealers get paid.

"A Harley?" I said. "What's your monthly payment on the truck and the bike?"

"Come on, Caroline. I don't want a math lesson. What do you think? Is it me?" he smiled eagerly.

"Honestly, it's your money, but I love it." I put my foot up on the trailer hitch. "Especially the towing package."

★

I pushed open the door to our apartment, calling out to Ali and the other girls inside, "You'll never guess who's coming to help me move the Avion."

"Caroline . . ." I couldn't see Ali, but I could tell from the teary sound of her voice that something was wrong. I rushed inside and found her slumped at the kitchen island, her face in her hands. "I'm done."

"What do you mean?" I asked, sliding into the seat next to her.

"It's medical. My hip."

During the intensive medical evaluations required for flight school, the Navy doctors discovered that Ali had a rare hip condition. While it didn't require immediate surgery, it could become debilitating over time.

"But you're in better shape than all of us. You're running more than five miles a day."

"They say it could be exacerbated by movements required in the cockpit." Her voice fell to a whisper. "I'm disqualified. I'll never be a Navy pilot."

With my best friend suddenly gone from flight school, I had even more reason to keep busy, so I turned my attention to finishing the Avion. As hoped for, the Minotaur helped me move the trailer to an open field near a hardware store—a far better place to work on her but still not a home.

"It's DEMOLITION DAY! The BEST day!" I chanted the phrase I'd heard so many times on home improvement shows as I began gutting the tin can down to her bones.

I cleared everything out, leaving only the beautiful blond maple cabinetry, working appliances, and bathtub. Finally with a clean slate to work from, I drafted my plan to phase the reconstruction, tackling major infrastructure items like plumbing and electrical first, and then moving on to flooring, building the bed, new countertops, and refinishing the cabinets.

Operating without electricity was tough, but it also kept me from overworking. Just before dawn each morning, I jumped into my Academy-issued sweats, twirled my hair into a messy bun, and worked until all my tools died, forcing me to let the batteries charge while I grabbed lunch and more supplies.

I was in a flooring shop, inspecting samples one morning, when I heard footsteps and the familiar words, "Excuse me, miss, may I help you?"

I turned around to see a guy in his early seventies, deeply tanned, slightly hunched over, with short, cropped hair. He had a clean look, and judging by his posture and direct line of sight, I assumed he was a former enlisted kind of guy. Probably did his twenty years, got out, worked at a flooring business, and eventually opened a store when his back and his knees gave out.

"Sure," I said, humoring him. "I'm looking to install flooring in my trailer."

"What kind?" he asked quickly.

"Nineteen sixty-nine Avion."

He nodded. "Warping is gonna be an issue. May I ask where the trailer is going to be parked?"

"Well, that's the thing," I said. "I don't quite have a place for it yet." I laughed in a mildly ditzy way.

"So what brings you down to Pensacola?" I saw his eyes running the same kind of math on me that I'd done on him—girl in her early twenties, just moved to Florida, either college student, military wife, or beach bum. "Husband in the Navy?" He came right out with it.

"Nope, no husband." I flashed my empty ring finger. "I'm here for flight school."

He made a face, surprised in pleasure. "Flight school?" he trailed off.

"Yes, sir."

"Looks like the students are a lot prettier than when I went through. I'm Bob Pappas," he said, extending his hand and smiling.

I would learn later we had both misjudged each other. Not only was Bob a former Marine officer, he'd also flown F-4 Phantoms in Vietnam. The conversation shifted from floors to flight school and the military. Soon we were sitting around drinking Diet Cokes and talking about the Navy until Bob paused and asked a question I'd hear again and again in the coming years.

"You ever think about going jets?"

★

Winter 2010; Perdido Key, Pensacola, FL

Eventually I found the perfect spot for the Avion, an RV park right on the water, in a nice part of Pensacola that catered to snowbirds. The only problem was it only accepted class A motorhomes, meaning the most luxurious class of RVs. But figuring it never hurts to ask, I brought the owner photographs of the Avion, proudly pointing out everything from the shiny rivets to new faux-hardwood floors and flowing drapes—all crafted by me.

"Yeah. Umm . . . well." The owner looked down at the pictures and up at my pleading face.

"You'll be the youngest resident with the oldest rig," she said, dipping her flamingo-shaped sunglasses, "but . . . I guess we'll let you in."

"Thank you." It was all I could do not to kiss her hands and happy dance across the lobby.

"It's only because you've got a classic rig . . . welcome to Perdido Key RV Resort."

After I was officially accepted into the neighborhood, the Minotaur once again tested out his tow hitch and dragged the tin can through the RV resort with a flock of aging snowbirds trailing behind us, as curious about the rig as they were about the two young people towing it. The Minotaur was clearly having trouble backing into my slip when an old sunbather came out from the shade of his gargantuan motor coach.

"Boy," he said, his skin weathered like a snakeskin boot and hair like white cotton balls. "If you need help there, let me know. Happy to show you how to park a rig like that."

"Thank you, sir, but I've got this," Minotaur responded through the window, gritting his teeth and grinding gears.

Mr. Snakeskin strolled up next to me, smelling sweet, like Tropicana sun oil. "So you're the aviator," he said. "My wife's told me about you."

"Oh, who's your wife?" I asked.

"You haven't met her yet," he said, "but she heard about you from one of the other girls."

He read my look.

"Yes, ma'am, hens will cluck. Name's Ken." He extended a tan hand. "Is Mr. Monster Truck your husband or boyfriend?" He nodded at the Minotaur, still spinning his tires in the sand.

"Just a friend with a truck."

"Uh huh," Ken said. "Good to know 'cause I got some grandsons I'd like to introduce you to. Good luck with this one," he said and headed back to his lawn chair, repositioning for maximum sun intensity before sitting down.

I downplayed it if asked, but I guess you can say things were going pretty well between the Minotaur and me. We'd taken some fun trips on weekends, and most days at sunset we met up to cook dinner or go out on the town. He was in charge of a group of Marines as they went through their training, so when he wasn't busy with his enlisted, he helped me work on the tin can. It was still confusing, though, stuck in a sort-of-friend, sort-of-relationship demilitarized zone. A lot of high fives and the occasional kiss, but regardless, I was glad to have the Minotaur in my life, especially since Ali had gone.

Later that first night I returned from base and there on the steps of the trailer was a casserole dish, tightly covered in tinfoil, a wooden bowl of salad, a Ziploc bag of dressing, and a simple note that read: *See you are working late. Enjoy! Your neighbors, Ken and Lydia*

★

At 0645 the next morning as the sun was rising, I formed up outside the schoolhouse on base with my new class of forty Navy ensigns, Marine second lieutenants, and French navy ensigns. The mood was tense, more than just your typical first-day nerves. Regulations had become so strict that any failure to pass the physical tests on the first day of API meant the student lost their spot in flight school. There was zero tolerance, so while the timed mile-and-a-half run was not a serious challenge, the consequences of failure meant we would not get our wings.

"You will complete three full laps," the instructor barked as we lined up on the trail. "*Do not* be the last one. *Do not* dawdle. This is your first test of many. Don't let us down."

Bang! A gunshot ripped through the otherwise still morning, and the forty of us set off, beginning the mile and a half at a near sprint. Soon, my lungs burned and calves seized as my feet slipped in the wet chips and sand. Rounding the bend on the final stretch, I saw the finish line and thought of Ali, how hard she had worked and how fast she had gone. *I'm not leaving.* I dug deep, pushing at full speed all the way to the end.

"Rogers, Montgomery, too slow!" the instructor screamed to a couple of guys who came in just after the cut-off time, huffing with their hands on their knees. "You're out."

Nice knowing you, I thought. It was as simple as that.

The next four weeks of API became a blur of early-morning workouts followed by full days of lectures and coursework, then studying at night. Because flight school was overloaded with students that year, the Navy raised the passing grade on academic tests from 80 to 92 percent. We constantly crammed in information—aerodynamics, aviation weather, aircraft systems, navigation, flight rules and regulations. The courses came fast and furious. An intense two or three days of lectures followed immediately by a written test. Attrition rates were high, and every few days, our class shrunk.

The Minotaur hadn't started his flight-school classes, but he would soon go through the exact API course a few weeks after me, so every night he studied with me to get ahead. We fell into a nice routine. After I left my study group, I drove back to the RV resort and there he waited for me on the gate of his truck, a stack of books beside him. One of these nights I arrived home and discovered the Minotaur was not alone. He had a female companion with him, staring at me with dreamy brown eyes. Her hair was thick, a beautiful mix of black and blond, and she looked like she weighed just under ten pounds.

"Kona, meet Caroline," he said. "She's a full German Shepherd. Pick of the litter."

"Hi," I whispered, patting her soft head and kissing her nose. I'd wanted a dog myself, but knowing I didn't have time or space in my life, I'd refrained. I took the puppy in my hands and looked up at Minotaur. "She's perfect."

"Try not to fall in love." He winked.

★

After four weeks of academic testing, only thirty some of the forty original members of our API class remained. We reported to a giant warehouse on the far side of the naval base and lined up.

"Here you go." The instructor dropped a bundle at my feet—two green and two tan flightsuits, the kind made famous in *Top Gun*. I was also issued a helmet, a massive bag of flight gear, and the official naval acceptance document making me responsible for the items. "Sign that."

I eyed the total at the bottom of the page. "Five thousand dollars?"

"Yep." The instructor nodded. "Don't lose anything."

I hefted the parachute bag onto my shoulder and headed to my car, trying to contain the giant smile spreading across my face. Back home, I stripped off my sticky shorts and slipped into my new flight suit, lacing my brown boots up tight. My helmet and aviators on, I opened the closet door and stared into the mirror hanging inside, almost unable to recognize the girl looking back at me. I twirled like a bride admiring her wedding dress. And it was a marriage of sorts. I had eloped and wedded myself to the Navy. *Caroline Johnson, official student naval flight officer.* My naval aviation career had finally begun.

Later that night, still in my flight suit, I met up with Minotaur, who teased me about how proud and excited I was to be in my official flight gear. But I knew he was a little jealous and couldn't wait until he could do the same in a few weeks. After a little reenactment of Maverick and Charlie but with the roles reversed, we left the Minotaur's house to head to the O'Club, or the Officers' Club, on base. Technically, as ensigns, we were officers and allowed to visit the Officers' Club, but this was the first time we were *actually* welcome.

Minotaur didn't yet have a flight suit, so he dressed for the occasion in crisp shorts and a button-down Vineyard Vines shirt, but I opted to stay in my newly earned flight suit, fully decked out with patches, and Ray-Bans propped on my head.

You could hear the party, the chatter spilling out into the parking lot. I took a slow, deep breath—part excitement, part nerves—as we stepped into the O'Club foyer replete with marble, fancy furniture, and chandeliers. I followed Minotaur into the bar, realizing that the rest of the Officers' Club was pretty much like any other midgrade restaurant—military decor, drinks, and pub food. But it was more than a bar. In an O'Club, surrounded by memorabilia of warriors and fallen heroes, the cheap beer, greasy burgers, and wilted salad you enjoy is earned by sacrifice and commitment to one's country and to one's fellow servicemen and women, making it all just taste better.

I slid into the seat next to an instructor who'd previously been terse and cool to me in class. "Stella," I said to the bartender, and as I sipped my cold beer, I felt the decades of naval tradition and culture wrap itself around me and lift me up.

"Cheers." The instructor turned to me and clinked my glass, acknowledging something far greater than a beer after work. We toasted an old way of life, and a new way of life for me. My head spinning, for once I didn't try to suppress my smile, and waited for my feet to touch the ground again.

CHAPTER SEVEN

★

February 2014; Embarked USS *George H.W. Bush*, middle of the Atlantic Ocean

During the first week of the cruise, in addition to making ourselves comfortable in our new digs, we settled into the boat routine. While onboard Mother, aviators operate on a totally different schedule than the ship's company. Sailors and officers assigned to the ship wake up and go to bed early, standing watch at all hours of the day and night. Aviators do just the opposite, getting up late and staying up late. Our Sailors worked twelve-hour shifts, half on days and half on nights, to make sure our aircraft were always ready to fly. The reason for our late schedule was that we flew most of our daily sorties (flights) at night, which is one of the greatest advantages the US Navy has over its foreign adversaries. The United States is one of the only countries in the world to dare to take off and land on aircraft carriers at night, and therefore, US naval aviators are night owls.

One of the biggest perks for aviators is that, per safety regs, we always get at least eight hours of uninterrupted sleep, and since we all stayed up late, most pilots and NFOs sleep in until ten, though dozing until the crack of noon was not unheard of in the Sharktank. I'm not normally an early riser, but on the boat I liked to get up at eight. I enjoyed having coffee by myself, where I could read the news online and answer emails in peace. Timing was critical, because each morning from eight to nine all of the ship's company performed cleaning station duty, and with four thousand Sailors thus occupied and away from their computers, I could use the extra bandwidth to get on the Google News mobile site.

My morning ritual included powering on the squadron laptop Taylor and I shared and opening as many news articles as I could in different tabs so they would load while I headed over to our in-room coffee bar to caffeinate. When I'd sit down to check emails, the articles would still . . . continue . . . to load. Internet at sea is slower than the original AOL dial-up, and when the news did load, it would be sans pictures, thanks to bandwidth restrictions and security measures. Additionally, all email was screened, so if you passed any sensitive information or violated any rules—sent a nude photo or used a trigger word like *bomb* or *Iraq*—your Internet privileges would be revoked for up to a month. My family and I always used code words for flagged topics and a cryptic number system for the dates I'd be in port.

After catching up on current events and email, when the cleaning stations ended at nine a.m., I'd head to the officers' gym—a tiny space about the size of a motel room—shared by three hundred officers. It had two treadmills, a stationary bike, an elliptical, three weight benches, two pull-up bars, and a set of free weights. My workout, like everything else on ship, had to be timed just right. If there were more than seven people in the gym at once, it was almost impossible to move around, let alone work out in the cramped space. Assuming I had space to work out, I'd shower and make it back to the Sharktank just as the other girls stirred in their racks.

As my roommates woke up and did the email shuffle, I studied for my master's in Administrative Leadership at the University of Oklahoma. I would end up completing half the coursework while on deployment, finishing the degree in about a year and a half, all while working and flying full-time. Of my eight squadronmates who started the program with me on deployment, I would be the only one to complete it. I'd like to believe this had to do with academic ability, but in truth, I give a lot of credit to getting up early, working late, and timing my days just right.

Like most aviators, my first meal of the day was lunch, then off to the Ready Room that all thirty-four officers in my squadron shared. There, I got down to business. Every day I wrote the squadron flight schedule (my ground job at the time), studied tactics, briefed, flew, debriefed, had dinner, maybe flew again, hung out, then went to midnight rations, or MIDRATS, followed by cowboy time—coming together after the events of the day, sitting around a big table and telling tall tales of our flights, life, tactics, or, really, anything.

Sometimes at the end of a long day, we would watch a movie in the Ready Room, usually a war movie chosen by one of the guys, something like *Lone Survivor.* Or the ladies and I would go back to the Sharktank, pull down the movie screen, and binge on *Downton Abbey* or old episodes of *Sex and the City.* Of course, seeing Carrie and Mr. Big up there on the screen, I'd start wondering about Minotaur in Afghanistan. He had joined his combat squadron when I joined mine, the Blacklions, and was already on his second overseas deployment. He'd checked in through email from time to time, but our contact had tapered off.

But it wasn't like I had much time to pine away for him. As we transited the Atlantic, we flew every few days in order to stay current. The goal of our flights at this early stage in deployment was to take off, climb up high to conserve gas, practice tactics, and land safely. We went through the switchology for dropping bombs, made sure we had enough gas, performed some basic maneuvering, checked our fuel some more, landed, and then did it all over again. When flying around the boat at sea, we knew fuel was life, dictating anything and everything we did in the flight. Some days there weren't any diverts to go to if we ran low, and the carrier wouldn't allow us to land outside the normal launch and recovery cycle if we got ourselves in extremis, so it was critical to always hawk the fuel gauge.

Out in the open ocean, en route to Europe, the seas often became incredibly choppy, complicating landings. On normal days, it was challenging enough trying to land on a flight deck while it drove away from us, but then we had to do so while the slippery deck heaved in the huge waves. If, for some reason, the conditions were just too severe to land on boat that day, protocol dictated we divert to a foreign country where we would land with little support. In anticipation of this, I always carried international charts and airfield diagrams for the runways nearby, as well as an international cell phone, credit card, my passport, a fresh change of underwear, and a toothbrush just in case my pilot and I got stuck for the night. When it came time to transit the Straits of Gibraltar, I half hoped an angry sea would force me to land in Spain where I could do a little shopping. Sadly, the Mediterranean calmed and my night in Palma de Mallorca had to wait.

CHAPTER EIGHT

★

May 15, 2010; NAS Pensacola,
Pensacola, FL

There are times in flight school during extreme physical training when being a girl actually helps, like during swimming qualifications. Women's bodies are more buoyant, given their relatively lower bone and muscle density and naturally higher percentage of body fat. Some of the men—particularly the ripped guys and quite often, the ripped Marines who come from inland states—struggled with the many swim qualifications we were required to complete in flight suits and survival gear. Particularly daunting was the mile swim in full flight gear. Of course many of these same boys started out in the fast lanes, but after a couple of laps, they wound up doggy paddling, kicking those who tried to pass with steel toe boots.

During the last two weeks of API, our first phase of flight school, the physical tests ratcheted up—parachuting, ejection-seat training, land survival, and a spatial-disorientation test affectionately called "the spin and puke" (a mechanical chair that spins around like a dreidel). And, of course, there was the dreaded helo dunker, which is exactly what it sounds like—a replica of a helicopter body that is submerged in a pool with students fully strapped in.

"It's just forty-five seconds of your life," the instructor said, half encouraging, half taunting before we dropped at 9.8 meters per second into a giant pool. As soon as the water rushed, the helo spun violently upside down and everything grew dark.

Forty-five seconds, forty-five seconds, I told myself until I felt the shaking stop. I grabbed my anchor handle with one hand and released my five-point harness with the other, but one of the belts got stuck. I fumbled for a few seconds, then finally broke free. My eyes burned with chlorine but I found my window and pulled my body free.

I counted my classmates bobbing in the water. *Seven. Three missing.* A few seconds later, another popped up, his helmet falling over his eyes as he gasped for air. Then another. Finally, after what felt like an eternity, our last classmate, a Marine, surfaced, breathing from the instructor's scuba hose.

With activities this intense, it was little wonder why API was so polarizing. Students either had the time of their lives or they were miserable and stressed, worrying that they were going to fail out. I guess I fell somewhere in the middle. While API wasn't the most challenging part of my career by any stretch, it also wasn't the most fun. For me, training was checking boxes, albeit important boxes, like which flares to use if your plane goes down. Flares come in two varieties—flares for the day and flares for the night. If you've gone down at night and need to send the signal for your rescue and you only have one shot at getting it right, it's critical that you know immediately which one is right. The Navy has made identifying the flares by feel very easy for us and easy to remember, too. The nighttime flares are ribbed for pleasure.

★

Summer 2010; NAS Pensacola, Pensacola, FL

In the military there is an inordinate amount of technical jargon and acronyms, especially in aviation. The terminology and the technical aspects of what we do can be incredibly confusing and almost impossible for someone outside our community to follow. Even for those of us who are in it, it sometimes sounds like we're speaking a different language. So, I will try to keep things simple and break down a few concepts.

First, while a student naval officer (SNFO) or pilot is in training, they move through a series of training squadrons, which basically amount to a collection of planes and instructors who teach certain skill sets. The Navy is very regimented and logical in its aviation training pipeline—all pilot and NFO students start at API to get their prerequisites knocked out, then they go to primary training squadrons and learn the basics of flying. After six months in primary, they go on to an intermediate training squadron, and finally an advanced training squadron where they'll eventually get their wings. This whole process lasts anywhere from one to two and a half years, and throughout all the phases of training, students are taught by both military and civilian contractor instructors. Each squadron has its own name and mascot. The first squadron I went to following API was VT-10 Wildcats, a primary training squadron. In this case, *V* refers to fixed-wing aircraft, *T* refers to training, and *10* is the number of the squadron.

Training squadrons are often known by fleet aviators as "clown squadrons," and it's easy to guess why. Painted bright orange and white, our training aircraft are easily spotted in the sky, as opposed to the sleek, gray planes flown in the fleet that are designed to blend in with the environmentals. All of my training would build toward moving from clown planes to fleet-gray warbirds, the training becoming more complex as I progressed through more challenging aircraft and assignments.

While there is little choice in the military and aviation community, students list their preferences for what they want to do, but they are just that, preferences. The only way you have a choice is by being number one in your class. But even then, if the desired slot wasn't available that week, the Navy would put that student where they needed them.

In my fledgling aviation career, I'd seen, studied, and even sat in the cockpit of Navy planes, but up to this point, all training on naval aircraft had been on the ground, like an intense coin-operated kiddie plane at the mall. When it was finally time to take this knowledge to the sky, I did so with instructors who could fail me, whether I was ranked number one or dead last. So number one meant one thing to me: choice.

My military instructors early in flight school were generally younger, active-duty officers, mostly lieutenants with five to eight years of flying experience, but the one thing they all had in common is that they were much saltier than me. And when we call someone salty, we mean extremely seasoned. They'd been there, done that, and gotten the T-shirt. They'd seen things, survived things, and so were generally unimpressed with everything, especially their students. Which was unfortunate for us, as we looked to the salty guides, hoping for favor like mortals before the gods.

At the time, I was the only girl in my class at VT-10. So out of twenty or so people, I was the only one with boobs. Well, there were some man boobs, but I was the only person with real ones, which meant there was no flying under the radar (pun intended). I have a natural tendency for cracking jokes, but I'd perfected a bulletproof facade since my time at the Academy, and because of that, it probably didn't seem like I worked as hard or cared as much as the rest. The further I went through the winnowing process, the more I'd come to understand that this was a man's game. Sure, a woman could play, but it was only natural for me to put some walls around myself, retreating to the safety of my cozy tin can. At that point, no one but Minotaur had even seen my home at the RV park, so they had no idea that I ate, breathed, and slept flight school. They did know I lived in a mobile home, and the whispers were that I was "trailer trash" living down by the river. And I guess I didn't go out of my way to correct them. I just kept my head down and tried to stay on top in the cutthroat aviation community.

On one of my earliest training flights, I was paired with one of the saltiest, most terrifying instructors, who I'll call Lieutenant Hardass. Salty as they come, Hardass embodied classic Navy pilot swagger, from the don't-give-a-damn stare, to restrained aggression, to the pair of Costa sunglasses dangling from Croakies around his neck. He projected Pensacola beach bum/former fighter pilot through and through.

We met to brief our flight, and I watched him ticking off judgments as I approached—*ditzy, tall, another set of self-loading baggage.*

I braced myself for a brutal day, and firmly reached my hand out to introduce myself. "Good morning, sir. Ensign Johnson."

He met my handshake with a strong grip and mumbled his name before starting in on the brief.

"Call sign for the day is KATT 11," he said, and then started rapid-firing questions to test my preparation for the flight. I could feel him trying to get through it as fast as he could, which wasn't abnormal. These instructors were so good, so seasoned, so experienced, that all of the nuanced technicalities of flying rolled off their tongues so smoothly that flight briefings became second nature.

Lieutenant Hardass concluded his spiel by pushing back from the table. "Questions?" he said.

"No, sir. Let's do this." I tried to show enthusiasm without overdoing it. Overdoing it was the worst. The instructors could see through contrived enthusiasm before the words even came out.

We went downstairs to the maintenance desk to sign for the plane, a process that involved dealing with retired Navy chiefs who'd become civilian contractors. For the most part, these are stout old guys, mechanics, and maintenance experts. Men that were past their military prime and it showed. If they camped out next to you at the beach, unfurling their Budweiser towels during your honeymoon, you'd probably consider moving a dozen yards away and upwind. The US government pays for the planes, we fly them, but make no mistake, the maintainers *own* the planes. They have the power to ground you or let planes stay broken just as fast as they can jump through a million hoops to make sure that you can fly that day. You can tell a lot about an aviator by how they treat maintainers.

"Mornin', Bobby," Hardass said to the first mechanic, an old grizzled Marine who looked like he'd rather be riding a Harley or draining beers. "You get your Bass boat out on the water, or are you just pretending to fish?"

Bobby laughed. "Hiya, boss." He wiped his hands on a rag and handed over a fat maintenance log. "Got everything ready for ya."

The warm reception Hardass got from the crew told me at the very least he was a respected pilot and a decent person. Maybe I'd misjudged him. We read the two-inch-thick book of records and noted any outstanding issues with our aircraft before LT Hardass signed the log and we proceeded to the paraloft to get dressed.

Dressing for a flight is something of an unspoken race—who can put their G-suit on or strap their harness faster. So of course LT Hardass entered the paraloft like Superman stepping into a phone booth, gear snapping on and zippers sliding shut. I tried to keep up, fingers flying, trying to keep my cool, even though inside I was going full speed and starting to sweat.

Hardass grabbed his raggedy well-worn helmet bag and stepped toward the door. "Ready?" he asked.

"Yes, sir."

He glanced back as I was a single step behind him, hefting my own helmet bag over my shoulder. "Huh," he said approvingly. "That's a surprise."

OK, maybe I'm not a total clown show, I thought, striding out to our plane.

Once we strapped into the cockpit, another race began as we blazed through our checklists.

"Cockpit checklist. Strapped in—uppers, lowers, legs complete," I relayed the checks, and he replied to my commands as we fell into an expeditious cadence.

When I was in the fleet and no longer a student, I wouldn't verbalize all the steps to start the aircraft, but during the rudimentary phase of training, I had to build the muscle memory.

Hardass couldn't see what I was doing, but he could hear it. The minutiae was tedious, but it would keep me alive someday.

"On/off, standby, check . . . check . . . check." We continued for five minutes. Once we were ready, instead of asking LT Hardass, I contacted the airfield's ground control. "Sherman Ground, KATT 11, Taxi Wildcat line to the active with Bravo."

Translated: "Hey, Navy Pensacola ground, my call sign is KATT 11, and we're trying to leave the VT-10 Wildcat parking area and would like to please taxi out to the runway. Oh, and by the way, we have the current weather information—bravo."

The plane lurched forward, and a surge of self-doubt rocked through me.

Should I have asked permission? I wondered, but then thought better of it. *Screw it. I'm going for gold on this flight. Let him tell me to ratchet it back.*

Outside of mandatory communications between the student and instructor, there isn't much talk because maintaining a sterile cockpit is critical. As a student, especially early on, this is challenging, because you are doing complicated things—sometimes for the first or second or third time—things that could kill you, or at minimum, get you kicked out of flight school. Sometimes you don't know if you are making a mistake, so your instinct is to ask first, and the students who ask are generally the ones who don't perform well. As the only girl in my class, I didn't want to fall behind and I certainly didn't want Hardass to think I was unsure.

After a myriad of pretakeoff checks I switched to tower and got clearance.

"KATT 11, cleared for takeoff." Before I knew it, the plane glided down the runway, and we were airborne.

During these instrument training flights, the objective is to fly point to point, to point, to point, to point, while making turns, climbing, descending, and performing complex calculations as we manually corrected headings for the wind, etc. These were all baby steps, but that day, they nearly overwhelmed me.

Equipped with a kneeboard and a whiz wheel, an aviation calculator with the technology of an abacus, I figured speed, crab angle, wind-corrected heading, and position while factoring in headwinds and tailwinds, other traffic in the sky, and any curve balls Hardass might throw at me. Numbers flying through my head, I felt like Rain Man, manually calculating over and over as our clown jet-raced across the sky. My body was physically restrained by eight straps in the ejection seat, but my mind was moving, pushed to its absolute limit. I tried not to acknowledge it, but I felt exhilarated, falling into the rhythm that developed between us.

Before I knew it, the flight was over, and the canopy popped open. Swampy Pensacola air fogged my visor and I swung my leg around to exit the cockpit, in perfect sync with Hardass, but only one of us could step out of our seats at a time.

"After you, sir," I said, deferring to his lead and gesturing at the wing.

Cockpit clear, I did the Super Mario Brothers hopping game to join him on the tarmac. I'd been listening to his voice and staring at the back of his helmet, but I hadn't seen his face or read his reaction in over an hour.

Once we crossed the tarmac, he lifted up his dark visor, his eyes far off, debating what to say. "You know." He put his hands on his hips and twisted his torso to crack his neck. "You definitely need to think about going jets."

I wobbled a bit. "I'm sorry, sir?"

"You heard me. I've never had that kind of flight with such a junior student. And I've never said this to an SNFO, but you really should think about going jets." He started toward the hangar and stopped one more time. "And I didn't even think *I* should go jets."

Feeling the ground underneath me, the sun on my neck. *Maybe he's right,* I let myself wonder for the first time. *Maybe I should go jets.*

★

Not every flight went as smoothly as my first ballet with Lieutenant Hardass. A few weeks later, with a little begging, I secured a spot on a flight trip with all senior students to Traverse City, the local airport near Grammy's cottage on Torch Lake. I started off uber confident, then found myself fourteen hours into the flight day, well past our regs, flying in bad weather and working my whiz wheel so hard I could smell burning plastic. A storm cell had moved in over the Traverse City airport, so air traffic control was stressing out, stacking all the inbound aircraft up in a holding pattern above the field while we waited for the weather to pass. I felt my mental edge slipping, but I still managed to correctly read back ATC's instructions and figure out our complicated technique to enter holding at sixteen thousand feet. I was furiously scrawling on my kneeboard, running our fuel numbers to see how long we had until we'd have to divert to our weather backup when my instructor, Lieutenant Cherry, started barking.

"Johnson, where are you back there? Taking a nap? You should be verbalizing the procedures for holding! We have to turn and you haven't even told me what our outbound heading is!"

This went on for another half hour until mercifully we were routed to land. We parked the jet and hopped onto the tarmac, and LT Cherry continued his lashing. "Johnson, what happened out there?" he said, lifting his visor so he could project his booming voice. "You weren't sharp. You weren't on at all. I never should have taken the risk of bringing along such a junior student."

"I'm sorry, sir," I said. "It's been a long day, and I'm spent—"

"Whooaaa." He raised his voice again so that everyone else could hear. "You need to man up, Johnson. Yeah, we went a little long, but this is nothing compared to what you're going to see in the fleet. In the fleet, this would be one of your easiest days." He threw the plane's covers and tie-downs at me. "If I'd known you couldn't hack it, there's no way I would've brought you along."

Instead of hurrying to put up the plane, I just stared at the tie-downs on the ground.

"Sir," I said calmly but with my fists clenched. "I was with it, all the way until the last thirty minutes. We're more than four hours past our crew day. Sorry I was dragging ass and my performance slipped, but I'm still learning. In order for me to do what you're doing, I have to work ten, maybe even a hundred times harder than you."

I could feel the eyes of the other students burning into my skin, but I was too fired up to stop. "It's unreasonable to think that you can break crew day in extremely challenging conditions and not expect that I would be exhausted. I know there are hard days out there, but this is not the fleet. I don't know, maybe I'm not cut out for this."

His eyes were on fire. "Johnson, that's the last thing I'd expected out of you. You better fix your attitude. And quick."

Fundamentally, he was right. I shouldn't have spoken back or made excuses. I'd volunteered for the trip and pushed myself too far. I wasn't prepared for the challenge. I was deeply disappointed in myself, and all the confidence that I'd felt after my flight with Hardass evaporated from me with the sweat on my drenched flight suit.

After the flight, the students split from the instructors. We found a liquor store and an open pizza joint, and eventually the dark gravel driveway that led to Grammy's cottage.

We ate the pizza and took our beers to the beach where the roar of a bonfire and a few local IPAs eased the day's tension and began to drown the voice in my head questioning my dream of flying in fighter jets.

"Damn, don't ever call Johnson out," one of the students joked, but no one laughed.

"Tomorrow's flight will be better," another one said, somewhat consoling.

"Yeah, for sure," I said, getting up to toss another log on the fire. Even among peers I was in constant deflect mode. I shifted the conversation to Torch Lake, telling stories about Grammy and the magic of summertime in Northern Michigan until the fire burned out. "I'll tidy up the beach and meet you guys inside. Take any room you want upstairs."

I hung down by the embers, trying to get my head straight, staring at the smooth, black surface of the lake glimmering in starlight. Back inside the cottage, I found the guys all studying the family photos on the wall.

"Hey, Johnson, that you there in the middle?" one of them asked. "Dude, where are the chicks in your family?"

"Eh, it's just me. I was the only girl growing up with all the boys up here," I said as I took a swig of my beer.

"Makes sense," he said after a few seconds and stumbled up the stairs. Following behind, I wondered *what* exactly made sense. That I was tough? That I knew how to play with the boys? That I was in the Navy?

We got up early the next morning for more training and when we got to the airport, LT Cherry had a surprise for us. Maybe he felt bad about the day before, but for whatever reason, instead of practicing our instrument procedures like we'd planned, we were going over to Wisconsin to do a flyby of Lambeau Field, home of the Green Bay Packers.

"Figured you guys worked on enough instruments yesterday, so today, we sightsee."

After the brief, I caught up with LT Cherry in the hangar. I'd been worried he was still mad about the night before, so I apologized for my lack of professionalism and attitude. He shrugged it off, responding with a quick concession of his own. An exchange of nods and a handshake later, both of us had moved on.

During the flight, LT Cherry took his apology one step further, and on the way back to Traverse City, he asked for the location of Grammy's cottage. I gave him a bearing and distance, thinking that he just wanted to look that way, but I was overjoyed when he radioed ATC.

"Request permission to proceed visually over Torch Lake."

As we turned toward the turquoise and blue gem in the distance, he dropped down low and flew down the length of the lake. Approaching the cottage, I pointed our beach out, and he rocked the wings as we buzzed the cabin.

LT Cherry keyed the ICS. "Work hard, play hard."

CHAPTER NINE

★

March 2014; Embarked USS *George H.W. Bush*, the Mediterranean

On an aircraft carrier, hundreds, if not thousands, of passageways wind through the decks above and below the waterline. Even during my nine-month prisonlike stay, I found it impossible to know everything that went on aboard Mother.

I primarily spent my time on the flight deck and the O3 level—the third deck up from the waterline, right below the flight deck. Aviators rarely left the upper levels of the ship other than to go downstairs to the lower wardroom for dinner or to visit the laundry facilities. You could compare a modern-day carrier to the transatlantic cruise ships of old, the levels of the ship representing social strata. In some ways, we were like *Downton Abbey* at sea, upstairs versus downstairs, and for the most part, we followed the hierarchy.

One afternoon Carolyn and I were walking across the hangar bay, lugging our bags of clean laundry when we saw a swarthy guy with a long beard flowing over a beaded necklace. He wore jeans and hipster glasses. "Who the hell is that?" I asked Carolyn, as the bearded man scurried past with a nervous nod. "He looks like a stowaway."

"One of those profs they brought along. I think Afghani tribal culture is his specialty." Unbeknownst to me and a few of the other nuggets, or aviators on their first deployment, before we left Virginia Beach, six civilian guests joined us on the *Bush*. They were professors and subject-matter experts onboard to educate us aviators on the countries around, over, and in which we'd be operating. When our lessons began, no topic was off limits—life and politics in the Middle East and Israel, the history of the region, the civil war in Syria, regional geopolitics, cultural nuances, even village life in Afghanistan and the leadership of Al-Qaeda. They were excellent teachers, and listening to them speak, I could almost taste the kebabs. Not only did I enjoy learning about the world in this way, but the lessons would also prove useful in a few short weeks.

In addition to these scholarly briefings, we refreshed our knowledge of Search Escape Rescue Evasion, or SERE, and devised personal survival and evasion plans in the event that we went down in combat. Helping us in this effort was a foreign ranking officer named Lorde. A commando turned Harrier pilot turned F/A-18 pilot, Lorde had been training with our squadron, and when it came time for SERE plans, he was the first to weigh in, detailing how he, a snake-eater turned pilot, would take on the Taliban.

"Right. Once we survive the ejection and parachute ride down, we will use hand signals to communicate with each other." He effectively took command of the Ready Room and with props—a machete and a model airplane—briefed us.

He held up a fist. "This signal, I'm sure we all know, means *stop*."

Another gesture. "And this means *get down* . . ." Not knowing we'd all recently refreshed our pistol qualifications, he then went on to passionately detail the marksmanship techniques we would use with our weapons, should we find ourselves on the ground with the enemy closing in. "Once we run out of bullets, we can use this." He brandished the oversized knife he carried when he flew. "And when our knives dull—which is to be expected when penetrating the enemy's flesh—we will use our bare hands." He held up his pasty hands. "Look at them. They are weapons you shan't ignore."

Will you also use your accent to kill any terrorists? I thought.

Along with the rest of my squadronmates, I suppressed smiles and snickers, knowing that if I was lucky enough to survive an ejection at twenty thousand feet in Afghanistan, I would likely break most bones in my body upon landing. No doubt, Lorde had a plan for this, too. And with his help, I could be a lethal pile of broken bones. Still, I had great affection for Lorde. He was a senior officer and quite a bit older than the rest of us. I didn't love flying with him, though, because he wasn't the best F/A-18 pilot, but on the ground he was fun and charming in a rakish, highly inappropriate way. What I liked most about Lorde was that he didn't give a flying fuck for proper military protocol and how things were supposed to be done. He only cared about performance and promotion, but unlike so many of the high-ranking American aviators who'd had a very regimented rise to the top, Lorde had followed a nonconventional career path, always scheming his way into his next dream job or amazing opportunity. And as a high-ranking officer, he'd had many perks and opportunities to do cool things.

Lorde also liked me. He liked me as a friend, a colleague, and more. A lot more.

We hadn't known each other all that long, but one afternoon when practicing airborne tanking, Lorde suddenly switched off the cockpit voice recorder. "You know, Caroline," he said over the communication system. "What about coming overseas?"

"What?" I asked.

"For your next job, come to my country as an exchange officer. That way, you could have it all—an awesome job in the city, and you could go to business school on the side."

"Well," I replied. "That does sound pretty awesome."

He went on. "I'll help you find an apartment. I'll leave my wife and daughter, and you and I . . ."

"Your what?" I cut him off. "Whoa, whoa, Lorde, stop. You have a family?"

"Naturally . . ."

His words hit like a punch to my solar plexus. Had any of my American squadron members propose I help break up a family midflight, I probably would have pulled the ejection handle, but since it was Lorde, I refrained, and rather than engage further, I flipped the cockpit recorders back on, forcing him to drop the subject.

Even if Lorde had been single, his offer was off the table because he was training with my squadron. Like most Jet Girls, I had a rule—don't date where you work. For one thing, it's dangerous to mix emotions and flying, literally. You cannot let a lover's quarrel cloud your head while you are in a jet. But more importantly, as a woman, I knew I had to be extremely careful with my reputation. It was hard enough to earn respect. I never wanted anyone to be able to say that my career success had to do with anything other than performance.

In addition to briefs from professors, more senior aviators, and Lorde, we also received in-person situation updates from our colleagues actively flying in Afghanistan. To do this, CAG and the admiral, our head bosses, arranged for the training officers to be flown off their carrier in the Gulf, across the Middle East and Europe, and eventually onto our boat in the Mediterranean. They then spent days updating us on the fight against the insurgents, briefing us on the latest skills and most effective tactics, and disclosing new pitfalls to look out for. Their real-world tips, tricks, and insight would become invaluable to us when we took over on station for them in less than a month.

Some of the less enjoyable but most important studying we did included reviewing the *Rules of Engagement*, or ROE, the classified document we accessed on our secret computer network or via printed versions—books that were more than three inches thick and kept in a safe. The ROE, written by lawyers in typical legalese, was so boring it always put me to sleep, but I knew the information was vital for how I would engage the enemy, what I could and couldn't do and when. As a WSO, I was responsible for our weaponry, and the last thing I wanted to do was harm a civilian or do something that would land me or my pilot in military jail. So with coffee in hand, I forced myself to memorize the bloody ROE, which most of us in the military believed, in truth, severely limited our ability to neutralize the enemy and save lives, which—there's no way around it—sometimes requires taking the lives of some bad guys. That is the essence of war, and as a Jet Girl, I was a warrior.

CHAPTER TEN

★

September 16, 2010; NAS Pensacola, Pensacola, FL;
Bay Minette Municipal Airport, AL

The next time you're taxiing across the tarmac, listening to the flight attendant drone on about seat belts and life vests, gaze out the window to the far side of the runways. One or more of the buildings you see will be the fixed-base operators, or FBOs, for the airport. The FBOs are contracted to handle refueling, routine maintenance, and other needs for all private and, in many cases, military aircraft.

From the largest airports to the smallest of airfields, each FBO has its own unique personality. Some are mom-and-pop operations, but increasingly larger companies are taking over this service, providing pilots' lounges and well-equipped lavatories, Wi-Fi and TVs, even comfortable couches for layovers during foul weather. The ones with heavy VIP traffic will have conference rooms, sleeping areas, and even small but tasty restaurants that showcase the local specialties. Needless to say, each place has its draw, and during flight school students and instructors negotiate their flight routes based on one very important data point—lunch.

"You guys in the mood for fried gator bites today? Or fried catfish fritters?" an instructor once asked.

"Is there a place where we can get both, sir?" a student responded.

"Both it is!"

Ummm, how about a salad? I wanted to say, but dared not upset the fry-cart.

The FBO in Bay Minette, Alabama, is famous in the aviator world for one thing above all else—golf-cart girls. Like the Hooters of airports, women in crop tops and teeny shorts gassed the planes and flirted with the pilots.

I appreciated the girls, too. It was nice to be around some estrogen again. I'd maybe compliment their earrings or lip gloss, and they'd tell me where I could get something similar. Climbing out of the trainer, I glanced over at the young girl about to gas up my plane, and did a double take when I realized she could have been me. My sweat-soaked flight suit and helmet hair aside, we looked alike, close in age and build. It seemed we both shared a penchant for makeup and long hair, and I couldn't help but think about how similar we were yet so different.

After all this time flying with the boys, I was ready for my first ever flight with a female Navy pilot. Her call sign was J. Lo, like the singer, and she was an incredible pilot, but because she wanted to start a family and be stationed with her husband (who was a Marine pilot at nearby Whiting Field), she opted to return to the squadron and teach. This was unusual, as normally the best pilots didn't go into primary training squadrons. They stayed flying fleet aircraft, grooming the next generation for war, sharpening their own tactics. Not only was J. Lo a top-rate pilot, she was both cool and high-performing, like a hot-girl version of LT Hardass. She wore her flight suit just like the rest of us, but somehow managed to rock it. Just enough makeup to look put together but not so much that you thought she was going clubbing. She wore a chic watch, high-fashion sunglasses on the tarmac, and Red Wing boots—a telltale sign that she was a seasoned Tailhooker. Tailhook aviators were known for upgrading their clunky flight boots to custom, beautifully constructed Red Wing aviator boots. Since the uniform is so standard, it's the little things that really made an individual in the military. And J. Lo, with her polished details, was exactly the type of aviator I wanted to be.

Our first flight came on a sunny Gulf Coast day. I was excited, not only to fly with a female pilot, but also to skip the fried gut-bomb the guys always sought out. We were going to get salads at the classiest FBO in the region, Million Air in Tallahassee.

We headed east from Pensacola at 220 knots (250 miles per hour), the crystalline gulf on our right, lush green forests on our left, and white beaches almost directly below us. Having experienced some less than perfect flights, I made sure I was well-rested and on the ball, providing every call and attending to every minuscule detail. I wanted to prove to J. Lo, and perhaps myself, that I was capable of performing at the highest level.

A good grade would also be nice, I thought, well aware that if I didn't get top of my class I might never fly in a fighter jet.

"Caroline," J. Lo said through our inner cockpit communications. "Don't take this the wrong way—you're doing great. Perfect, actually. But can we please just dial it back and enjoy the flight?"

I rocked back in my seat, stunned. Not only did she not want to hear my instrument procedures verbalized like the other instructors, but she'd actually called me by my first name—two things that hadn't happened before.

"I'm sorry, ma'am?" I asked.

"Listen, you're not in trouble and you shouldn't be sorry for doing your job, but can we just cut out these flight school games? Let's just relax. Look out the window."

And I looked.

"Can you believe we get to do this?" she said with a sigh. "I have a two-year-old at home, so I'm trying to escape that chaos. He's with the babysitter, and I'm so happy to get away for the day. All I want to do is fly this amazing machine, enjoy the sun and this beautiful view."

The plane droned on, and we sat in silence, gliding through the air. "So what do you want for lunch?" she asked after a few minutes. "We should call ahead so it's ready when we land."

Genius, I thought. Never had any of my instructors called ahead so we could enjoy our lunch without waiting in line, but J. Lo was schooling me in the Jet Girl way.

I dialed up the radio to the FBO's common frequency. "Yes, ma'am, KATT 61, tail N323, we'd like two chicken caesar salads and two Arnold Palmers. Be there in thirty minutes."

We landed and taxied off the runway, approaching the beautiful Million Air building, where there were probably ten huge private jets parked in front and five T-6s parked off to the side. It was a big University of Florida football weekend, so important alumni and NFL scouts were all in town and got priority parking on the ramp. J. Lo started to park next to our Navy counterparts, but the lineman recognized our tail number and waved us in between two big private jets, then actually rolled a red carpet from our wing to the golf cart he had waiting.

J. Lo and I hopped down and peeled off our gear. I shook out my helmet hair and threw it up into a bun, relieved when I looked over to see J. Lo doing the same. I realized in the silent race I'd had with the other instructors, they got to skip this step, never having to juggle a helmet bag while pulling stray hairs into a regulation updo.

We took to the red carpet like starlets at a film premiere and jumped in the golf cart. J. Lo handed our driver the government credit card.

"Mind topping her off?" She shot him a bright, genuine smile.

Inside the lobby, we bypassed the line of boys waiting to order. The staff handed us our salads and drinks, we thanked them for the meal, and they thanked us for our service. Once we were seated in a quiet corner, I couldn't keep quiet any longer.

"J. Lo," I said, leaning into the table. "That whole sequence was incredible. The faces on the other instructors—priceless. How often do you come to Tallahassee? Do they always give you a real red-carpet service?"

"I don't make it over here too much with my baby and all, but it's simple, really. I just try to treat everyone with respect. So many pilots think they're entitled to premier service because they buy thousands of dollars' worth of gas. Big deal." She paused, quietly crunching her romaine. "But if you're kind and acknowledge people, they go out of their way to take care of you."

Feeling around in our lunch bag for more dressing, I pulled out two decadent fudge brownies and a note. *J. Lo, great to see you again! Come back soon!*

J. Lo was both a mom and a rock star, and everyone knew it. After we debriefed our flight and talked about our plan for the leg home, we did what typical ladies would do if they were out lunching with a girlfriend—we talked about family, about men. In this male-dominated business, it felt amazing to connect with another female. J. Lo showed me that you can be a badass tactical pilot, and still have a life, a family. You can retain your grace and humility in the world of naval aviation that's all too often hyper-masculine and ego-driven.

On the way back to Pensacola, she took me on a beach tour, naming the gorgeous white-sand beaches, telling me which were the best. She showed me her family's beach house, and pointed out bars and restaurants I should try with the Minotaur.

As we neared the airport, I got to thinking about instructors and how wrong I'd been about some of them. J. Lo and LT Hardass and others like them might have been testing us, but ultimately, they wanted us to be part of the naval aviation team. They wanted us to have fun and to share in the cool things they were doing. They weren't just trying to do the best, be the best, get ahead. For many of them, it was more than that—they loved this job, this life, this community. I think they even loved us.

After the flight, I found out the grade I received that day was above MIF—the equivalent of an A+. J. Lo had shown me what it meant to be a Jet Girl, and from that point on, it was all I wanted.

As primary flight school ended and the class ranks were announced, I sat in the crowd of my peers, daydreaming about what J. Lo had said. *Can you believe we get to do this?* I was picturing the perfectly clear sky and emerald water beneath us and nearly missed it when they called my name. "Please congratulate Ensign Caroline Johnson," one of my instructors said to the crowd, "for finishing primary flight training as top grad."

CHAPTER ELEVEN

★

August 2014; Embarked USS *George H.W. Bush*, North Arabian Gulf

Running down the starboard side of the carrier was a passageway known as the Blue Tile, and predictably, the flooring in it is made of blue tile. Shielded on both sides by blue curtains bearing admiral's stars, the line and the demarcation were far more significant. No one was supposed to cross into the Blue Tile unless they were an officer on official business or an officer ranked O5 or higher, meaning commander and above. While most high-ranking naval officers worked their whole lives to get on the distinguished side of the tile, to enjoy the segregation and distinction that came with hierarchy, a few spent their time crossing it and welcoming others to do the same. And so even though I was an O3, a lieutenant, rarely with official business, I traipsed down the Blue Tile many days, without fear of repercussion, on my way to say hello to one of the finest leaders in the Navy.

Admiral Bullet was the most senior aviator in the Navy still flying F/A-18s. He'd been flying for thirty-three years and had logged almost a thousand traps, or carrier-arrested landings. Even though flying is principally a young person's game, Bullet had stayed in it. The jet demands physical strength, stamina, and coordination, and it requires constant study and practice to stay current on rapidly evolving technology and tactics. To compensate for this danger, Bullet would only fly two-seat Super Hornets, even though previously in his career he had flown mostly single-seat jets.

In training and in combat, pairings between pilots and WSOs are constantly rotated. Some days I was crewed up with pilots who were fresh out of training at VFA-106, the F/A-18 replacement squadron. On other days, I'd be paired with the admiral. This constant rotation kept people standardized, ensuring that no two pilots and WSOs got too accustomed to flying with one another because they cycled through every member in the squadron.

Of my handful of favorite pilots, Bullet was the one I was most happy to fly with. But hopping in a jet with an admiral was not something a WSO could just sign up for. The squadron commanding officer or the admiral's aide either chose for him or the admiral chose his backseater himself. In the beginning, he only flew with high-ranking, senior WSOs, but after Bullet and I flew together for the first time, I found myself scheduled with him quite often.

The admiral loved to fly, literally lifting himself from his jam-packed schedule of meetings, teleconferences, and emails that often stretched from five or six every morning until ten or eleven at night. He tried to fly two or three times a week throughout deployment, but he always stayed close to the boat, meaning he never flew into combat. Throughout his three decades of flying, he'd flown in combat hundreds of times, but once he became an admiral, Bullet left the in-country flights to the young bucks. This choice didn't have to do with the danger, but with endurance. The average combat mission is six and a half hours, plus the hour minimum you spend before and after you land. Eight and a half hours was too long for a man in his late fifties to spend crammed in an incredibly uncomfortable seat, wearing fifty pounds of gear, with bags and charts shoved in all around him.

"Look, I'm old, Dutch. I want you there to save us if I mess up or miss something," he said to me one afternoon. It was an absurd statement, as Bullet was one of the best, most methodical, and predictable pilots I'd ever flown with, but it was still good to hear.

I was very fortunate to spend time with Bullet outside the jet. Escaping his office in the Blue Tile, he'd come down to our Ready Room for his flight brief and to hang out before and after his flight. I prepared our mission cards and products for the flight, handing him the stack of pertinent data he needed for the day. Protocol dictated that all aviators, regardless of rank, must brief the safety rules and mission objectives for the day, and since Bullet enjoyed talking through the flights, I always let him do it. His briefings were quick but thorough, covering admin, tactical admin, emergencies, training rules, and finally, the mission.

As was typical in the fighter community, it was business first, then we'd shift to topics he deemed more important. But unlike other aviators who tended to talk about themselves, he'd always start by asking about me. He was passionate about leadership, and knowing I was hard at work on my master's, we often talked about leadership theory. Bullet always had great book recommendations. I'd write down new reads for my mom to send me. Often the conversations delved into other topics, including our families, and a few times he even sought my advice.

"Dutch, need to ask you something," he said as we sat in the Ready Room one morning. "My daughter, she's twenty-seven, like you. You wouldn't believe it with a father like me, but she's actually beautiful and smart as a whip."

"Yes, sir, I've seen your wife, and I'm sure she got the best of both of you," I joked.

"Yeah, yeah. Well, she's also outdoorsy. I really don't know where she got that from, but she's moved up to Alaska and got a great job as a producer at a news station."

"Sounds like she's doing awesome," I said, wondering what this had to do with me.

"Well, here comes the tough part, Dutch. I called her the other day, and she had some news. She's dating a guy . . . a man she met at fly-fishing class." He raised his eyebrows skeptically.

"Well, sir," I said. "Let me guess. You're wanting to check this guy out." I laughed, but my heart broke a little for him. Bullet was a family man. He clearly loved his daughter like crazy, but he loved the Navy, too. In thirty-three years, he'd already sacrificed so much, and here he was, continuing to put his family on a back burner while he served his country. Here was an admiral, responsible for seventy-five hundred people aboard twelve different ships, and untold billions of dollars. In charge of the most powerful array of naval and aviation forces the world had ever seen, and yet, he was powerless to protect his daughter from a guy who could be a sleazeball.

"So, what do I do, Dutch? Not like I can just drop everything and go," he said, turning to look at me. "Have my aide stalk him on Facebook? Fly my wife out to Alaska so she can check him out?"

Still a little unsure why he was sharing this with me, I proceeded cautiously. "Well, knowing your daughter is a product of you and Ellen, she probably has high standards. I'll bet the guy's down to earth, if she met him at fishing class," I said, thinking of the Minotaur. "And if it's the wrong guy, she'll know in time. These things sort themselves out."

I hoped these words were true for her, and also for myself, for the love I'd given up. For a guy somewhere in Afghanistan who hadn't emailed in months. "The only thing you have to check on is if he has a job." I winked.

Since we flew so often together, I'd started to learn Bullet's quirks. For one thing, he liked to take a detour on the boat en route to the jet. He knew every nook and cranny of his beloved *Bush,* strolling its deck like a master gardener might stroll through one of his flower beds in full bloom.

"See that right there?" he said once, pointing to the windsock. "When the *Bush* was designed, all of the wind readings were digitally displayed on the bridge for the officers driving the ship, so there was no need for a windsock. But during her first sea trials, we discovered that when trying to drive the ship directly into the wind to land ships onboard, all the technology in the world wasn't as effective as a cone of fabric blowing in the breeze." Bullet explained how he tried to get the ship builders to add an aftermarket windsock, but the price for this addition was astronomical, so instead he asked one of his admiral buddies if he could take the windsock off a decommissioned carrier that was now a museum in Texas. And that's exactly what he did.

Not only did Bullet know the ship intimately, he also knew its crew. Or more specifically, the roles each person performed. Like the transformational leader he was, he knew how to make each of the thousands of people onboard feel that they were special and that their contribution was unique.

"This way," Bullet said one afternoon, taking me to the very back corner of the flight deck where the arresting cables were controlled. Once an aircraft lands, it pulls the cable out. The cable itself is then disconnected from the jet's tailhook and is manually retracted back in place in order to catch the next jet coming in.

"I'd always thought it was an automatic system that functioned without much human oversight," I told him after taking the tour.

"Nope," Bullet corrected. "Let's take you up and let you see for yourself."

Geared up, we climbed onto the flight deck, weaving our way through the parked jets. When a jet on final approach touched down, I sprinted after Bullet as he ducked and ran for the back corner where an enlisted guy was standing. I had never noticed this Sailor before and was shocked that someone would stand so close to the runway for every single landing.

We scurried down a three-foot ladder, and Bullet greeted the Sailor with a pat on the shoulder. "Hello, there."

The kid turned, and seeing two aviators in front of him and then the stars on the admiral's shoulders, his eyes visibly expanded in his protective goggles.

"Sir, uh," he said. "Can I help you?"

As he stammered, I noticed the babylike freshness of the kid's face, and I could tell by the markings on his long-sleeved flight-deck jersey he was a third-class petty officer, an enlisted Sailor with the pay grade of E-4, who had to be no more than nineteen years old. He fumbled for words, but he was so well-trained that he continued to do his job with the joystick in his hand.

"Petty Officer," the admiral said. "Do you mind showing Dutch here how you retract the arresting gear back in battery?"

"Uh, yes, sir," he said. "Happy to."

The kid started his demonstration, and the admiral winked. "Pretty cool, huh?" he said.

It wasn't just cool, it was a critically important job, and as we listened to the petty officer dive into a five-minute detailed explanation about the arresting cable and the joystick, I watched the kid grow taller, exponentially more confident of his knowledge and the important function he served in the Navy.

After the spiel was over, Bullet turned to the guy and shook his hand. "Keep up the hard work, buddy. You're making a difference. We wouldn't be able to fly without you."

In that moment, I realized that this was the kind of power I wanted, the kind of power that could become addicting. Not lording your strength over others to make them feel weak, but actually showing people, those who might otherwise be blind to it, the worth in themselves.

"Sir," I told Bullet as we headed back to our plane. "I bet that after his shift, that kid is going to email his family and friends and tell them the strike group admiral personally commended him on a job well done."

"I hope he just knows how important he is." Bullet looked off toward the horizon, and nodded.

Bullets I Learned from Admiral Bullet

Lessons on Leadership

- Empower and support. If you want to recruit, hone, and retain the best, most talented people, you must motivate them by empowering and supporting their efforts to succeed at work and in their personal lives.

- Always accountable. Always hold yourself accountable and provide accountability for failure and more importantly the success of others. Recognize your people's accomplishments and celebrate them.

- Courage equals character. Courageous leaders act with unimpeachable character at all times.

- Relationships rule everything. Leadership is about developing relationships with your team—through expressing humility as well as vulnerability and strengthening with trust.

- A little empathy goes a long way. Great leaders take care to walk in the shoes of others (subordinates, higher-ups, enemies) and try to understand the battles they are fighting.

- Appreciate small wins. Leadership is about the little things, praising people for a job well done.

- Listen and convey with actions and words. Communication is about hearing as much as it is about talking—then when you do share, remember it's not just the words but the tone, body language, and the content.

- The higher you climb, the more you need to pay attention to your base. Even when you reach the top of the totem pole, don't forget about those at the bottom—as a matter of fact, ask those people at the base what they think; if you've built rapport and trust they will give you key insights about the health of your organization.

- Be aware of bias. We all have biases; to ignore them, or try to pretend they don't exist, is fatal to your leadership ability and toxic to your organization. Be aware of the biases you and others on your team possess—openly acknowledge and actively work to overcome those biases.

CHAPTER TWELVE

★

Spring 2011; NAS Pensacola, Pensacola, FL

Because I was top of my class, I got to put in my preference for what I wanted to fly after primary training: P-3s (which my brother flew), E-6s (big communications planes), or Tailhook (E-2s, E/A-18s, F/A-18s). There was no question in my mind what I would choose. I wanted to fly aircraft that landed on aircraft carriers. I wanted Tailhook.

While on the Tailhook track, our challenges intensified. Formation flying and acrobatics proved particularly difficult since we had to learn it on paper before we put it into practice. Grasping 3-D maneuvers in space, in relation to other aircraft, all moving forward at the same time, made our heads spin. And the descriptions in the Navy training manual were ludicrously abstract—hundreds of pages described physical maneuvers yet only a handful of simplified diagrams showed them in context.

I read aloud to our study group of Navy classmates, Marines, and a Middle Eastern student in training with us. "Okay, does this make sense to anyone? The wingman uses angle of bank to fly along the forty-five-degree bearing as the distance between the two aircraft is closed. Through the continual manipulation of this principle, the joining aircraft is able to maintain a steady bearing line as it approaches the lead aircraft . . ."

"Johnson, I have no clue what the hell they're saying." Matt, a Marine classmate, slammed his book shut. "My head hurts. I need a drink."

"Me too," I said, "but the test is tomorrow, and I don't think any of us could even re-create this stuff with . . ." My voice trailed off as an idea formed. "Let's get beers. And bikes. And try the maneuvers on bikes."

"I like the beer part," Matt said.

"Great. I have some. Meet me here." I jotted an address down. "And bring bikes or skateboards if you have them."

"Where's this?" Matt asked, glancing at my address, but halfway to the liquor store in his mind. "You live in a trailer park?"

"A class-A RV park actually," I said. "There's a big, open parking lot where we can practice. And Matt, don't you still live with your mom?" I asked, giving it right back.

"Last one there buys tomorrow's coffees."

As my classmates arrived at the Avion with bikes, skateboards, and six packs of beer, I could see hesitation on a few of their faces.

"Trailer?" Anwar, our classmate from the Middle East, said skeptically as he stepped up to the tin can and peered in. "How can a woman live in a trailer? And with this . . . little rat?" Anwar pointed to my hamster, Wizzo, running on her wheel by the window.

"Her name is Wizzo. She keeps me company." I dropped a few food pellets into her cage.

"Really, you didn't want a real pet instead? Like a dog, maybe?" Anwar said sarcastically.

He didn't know the half of it. I would have loved to have had a dog, but with flight school, I worked and traveled way too much to take care of such a high-maintenance pet. I had gotten Wizzo on a lonely weekend after Minotaur and I had broken up over a night of questionable behavior. I named her after my dream job—F/A-18 Weapon Systems Officer, or Wizzo.

Matt, the Marine, was last to enter. "Holy shit, Johnson. All this time we thought you were in some shabby little mobile home. For God's sake, look at these countertops. This is a princess palace."

"Thanks," I said. "My Grammy would be glad to hear that. Now, enough talking. Let's see if we can figure this out."

In the golden light of a Florida afternoon, we rode, one bike in lead and one bike behind the lead, in formation, weaving around my trailer park. It was the first time we'd actually had wingmen. I pedaled briskly, calling to my wing, "All right, take spread . . . back up to parade now . . . all right, get on bearing line, a little up now, all right, looks just right. Let's try a breakup and rendezvous now . . ."

As we rode, the concepts that proved so elusive on paper now became clear as we fell into sync. And quoting *Top Gun,* my lead called back to me laughing, "You can be my wingman anytime."

As our maneuvers improved and the beer drinking increased, a few crashes ensued.

"Hey, guys, who feels like a dip?" I said as heads turned toward the pale blue Perdido Key RV Resort swimming pool twinkling on the other side of the park. A few of the guys called their wives over, and our study session turned into a pool party. Nothing crazy, just drinking a few cold ones, telling jokes, enjoying a hot Florida night.

Head tilted back in full recline, Anwar floated over on a giant pink flamingo floatie, sunglasses on, even though the sun had set. "Caroline," he said. "I really do like this trailer living. I will miss this when I'm back home."

"I'm sure you'll miss a lot of things—the beers, pepperoni pizza, young Pensacola girls?" I raised my eyebrows, smiling.

Anwar laughed, fingers dipping in the water. "For the rest of my life."

"So," I asked him. "When you go back, will you marry?"

"Next year, I will probably marry my first wife. And in two or three years, I will probably marry my second."

"Geez, Anwar, how many wives do you think you'll have?"

"Two or three," he answered, without a trace of irony. "Caroline, tomorrow will you help me with the syllabus?"

Anwar had changed so much since we started primary flight training. Like most of the Middle Easterners, initially he wouldn't talk to me because women in his culture were not respected at work. But that all changed as soon as he realized I could help him.

"Sure," I said. "Happy to help."

★

Training progressed rapidly from flying the T-6A Texan IIs—the small, fast planes with glass canopies and front- and back-seat configurations—to T-39 Sabreliners, little Learjets with degraded F-16 radars. Fairly nimble and high-performing like the T-6's, Sabreliners had one distinct difference—they were *jets*.

Pensacola flight traffic is a five-ring circus. Due to the amount of military training, the airspace is filled with the same amount of planes as airports in New York City, but most of the aviators flying around Pensacola are just learning. In many cases, students have only been in the air a few weeks or months. Add to this volume of air traffic and the variation in flying tactics (everything from basic turns to performing aerobatics to elaborate war games with multiple aircraft), and being in the sky around the Florida Panhandle is simply dangerous. Flying with safe, experienced pilots is critical.

The best of the best instructors are known as the Grey Eagles, retired Marine and Navy pilots, many of whom are Vietnam veterans and former A-6 Intruder or F-4 Phantom pilots who simply love to fly every day and want to teach the next generation of aviators. They look like a geriatric baseball team but have hundreds of thousands of hours of flying time and hundreds of years of combined experience as military aviators, flying the squirliest, most dangerous planes the US owned. They had swagger, superior skills of the saltiest pilots, plus the humility that comes from extensive combat experience, including surviving ejections or having been shot down.

Flying with Grey Eagles was unlike anything else in flight training. For one thing, their purpose was to prepare us for a dangerous occupation more than it was to weed us out. In the air, it was businesslike and controlled, but they still knew how to push it. Of all the complex maneuvers we performed, my favorite was high-speed, low-level flying. We navigated solely off of ground features whizzing by—power lines, TV towers, interstates, mountains, and towns. So low, it felt like we were in a video game, our wings nearly clipping tall trees. But it was no video game, and all we had to do to remind ourselves of that was look at the scarred mountainside, a barren patch of dirt stretching the span of a football field, where ROKT 21's last flight ended only ten months prior. Three flight students, one Navy instructor, and a Grey Eagle had perished in the crash.

At the beginning of flight school, I'd been made to do the cursory, morbid tasks of signing a waiver in case of my death and writing a letter to my parents. While I had girlfriends back home planning weddings, I made my funeral arrangements at the ripe old age of twenty-three. I looked into doubling my life insurance, even though I didn't have more than a hamster and my parents as beneficiaries. On paper, the dangers felt very abstract, but they came to mean something altogether different at five hundred feet off the ground, going 350 miles per hour in a jet.

★

After completing intermediate flight training and again finishing number one in my class, I had to report to Lemoore, California, for centrifuge training. In Lemoore, you show up to a giant, hangarlike building in the middle of Nowhere, California, and there, in the center of a cavernous room, you are strapped into a porta potty–sized chamber suspended on the end of an extended gimbal rotating around an axis. You sit and wait to feel the spinning and the Gs coming on.

The goal—exerting extreme g-forces on the body—is to teach us how to counteract the blood leaving our brain and vital organs and pooling in our feet. This same sensation in the jet is what causes you to pass out. We had to train our bodies, learning to sequentially flex and clench all our muscles, from toes, to calves, to glutes on upward to keep the blood where the body needed it. Being spun around in a box with no visual reference, strapped to a chair like a monkey getting sling-shot into space, naturally induces nausea.

When my turn came, instead of focusing on not passing out or puking, I couldn't stop thinking about how my life had become the centrifuge, the Navy spinning it while all I could do was grit and resist the forces. I thought about Minotaur back in Florida and how soon he'd be finding out where he was going for his next round of training, painfully aware that the relationship we were shakily building together might not weather distance. I clenched my body, unsure if I had been inside the chamber ten minutes or ten hours, but I knew if I passed each of the four different profiles the first time, I'd only have to spend fifteen minutes spinning.

Finally the centrifuge slowed, the door creaked open, and an instructor had to help me out of the chamber because my legs were too wobbly to make it down the stairs. Luckily, the night before, I'd decided to rein in the margaritas and Mexican dinner whereas several in my training group enjoyed their tequila and nachos twice. Two of them even blacked out.

Conflicting thoughts tumbled around in my head on my trip back from centrifuge training, like emotionally I was still in a spin. Life in the military, in general, is difficult on relationships. There is no planning things out. You live and die by the schedule the Navy makes for you, and by and large, the Navy doesn't care if you get engaged, or have a date planned, or have a child. Well, if you are getting married, they will try to make an exception, but ninety-nine times out of one hundred, nuptials are scheduled according to Mother Navy's plan. When orders come down, in a matter of days, or sometimes even less, you are forced to move on to another phase of your life, often separated from friends and family. When service members are deployed overseas or must change duty stations stateside, the Navy doesn't always provide resources for the families to come along.

This uncertainty helped to make life with the Minotaur both thrilling and annoying. We never could reliably plan to be together, since our flight schedules were only written twenty-four hours out. Though we never discussed it, within a moment's notice one of us, or both, could be sent to the other side of the country or could die in the line of duty. This ephemerality forced us to live moment to moment, trying to suck as much out of that moment as we possibly could.

I also found it exquisitely annoying when I was hoping to enjoy one of those sweet life moments with my boyfriend, but because we were around his friends, he'd call me "buddy" or "kid." As in "Hey, buddy, whatcha doin'? Why don't you come over and sit next to me, kid?" Every time I heard those words, I wanted to punch him. And no doubt, he would have preferred a punch in the face rather than give way to openly showing affection or having a conversation about our relationship or discussing life plans. Not that talking about feelings was something I wanted to do, either. I'm pretty stoic, but compared to that guy, I'm like a tearful guest on *Oprah*.

As the plane touched down in Pensacola, Minotaur was there to greet me. In the airport he strode toward me with a cockeyed grin.

"She's back." He hugged me tight, smelling of salt and deodorant.

"Been swimming?" I asked.

"Paddleboarding." He took my bag from my shoulder. "So where do you want to eat—and pick somewhere nice 'cause we're celebrating." His face, normally as chipper as a statue on Easter Island, was near giddy with excitement.

"What's the occasion?" I asked, coyly.

"I got Whiting." He opened the passenger door of his truck and motioned for me to climb in. "I'm staying right here."

"Oh" was all I could say, but my smile said it all. "You happy?" I asked.

Kona, likely feeling the excitement in the cab of the truck, crawled into my lap, panting and glancing back and forth between us.

"Yeah," he said, patting her head. "Kona's gotten used to being a beach pup."

And suddenly I was spinning again, knowing that this had bought us time. He would be at the local helicopter training base, Whiting Field, until he finished flight school, which meant we'd be together until I finished flight training. It also meant neither of us had to confront what we would do if the military separated us. All those questions remained unexamined, at least openly, as we literally drove into a pink sunset.

★

When I showed up for VT-86, I knew I needed to level up, but the amount of work the Navy threw at us surprised me. And that was exactly the point. To overwhelm us. There was another surprise waiting for me on the first day of class.

As usual, I arrived early, coffee in hand, hair damp from a shower, endorphins still pumping through me from my morning jog. I took my usual place at the middle center of the class with my dog-eared notebook open on the desk when something unusual breezed past me.

Was that a . . . bun? Yep. I could smell her shampoo as she took the seat right in front of me. I stared in utter disbelief. For the first time in a very long time, I wasn't the only girl.

Our instructor blazed into the room, firing off questions and scribbling on the chalkboard. Before he even paused to ask a question, her hand shot up like one of those mechanical moles Craig and I used to whack at the arcade. Not only was she female, she was smart, and suddenly I smelled more than the aroma of Pantene in the air. I smelled winds of change.

Her name was Andrea, and I knew an old western shootout lay ahead of us. Andrea commissioned through NROTC at Northwestern University and then, like me, went to flight school as a SNFO. Until that moment, we'd only really known of each other. In a number of ways, she was my opposite: short, petite, and dark, with a deep, sexy radio voice, while I was tall and fair, my speaking tone bubbly and higher-pitched under the Gs of the jet. I'd come to find out later that like Ali, Andrea was also the product of a hardscrabble childhood. She grew up with a single mom, while I had parents whose support would have rivaled June and Ward Cleaver. But we were also cut from the same mold: hardworking, driven, and competitive.

Now, looking back at the experience, the most shocking part of having another girl in class with me was not the fact that it took almost a year and a half of flight school for that to happen, but that—and I say this in all humility—she was the only person to truly threaten my position as number one.

Andrea and I both knew we were never going to be best friends, though we quickly realized we made an amazing pair, sharpening each other at every turn. We competed in everything—in class, in the air, out partying in town, and even in who had the best boyfriend—Minotaur, or her scrawny but smart SNFO boy toy who was a class ahead of us in training. Everything became an unspoken battle, each of us trading first and second place over and over again. It was like the rest of the class disappeared, and we were in a knock-down, drag-out, albeit unspoken, contest.

The hypercompetitive environment was nothing new. Put twenty high-performing guys in together, and the level of competition rose, as did the stakes. Add a female to that same mix, so there were nineteen guys and one girl in the room, and the stakes shot even higher. No one wanted to lose to that girl. Coming in second place to a chick who lived in a floral-accented trailer and giggled at Hugh Grant rom-coms at night was just as bad as coming in dead last.

As flight school progressed to advanced training and winnowed all three intermediate classes down to just fifteen, I realized I'd become addicted to beating out the guys. So much so that I would do anything—study harder, stay up later, practice longer in the sims—to reach that number-one position. The real secret to being female in this rough-and-tumble man's world is that when you get to the top, it's so freaking fun, and so rewarding, that you'll be damned if you slip down through the ranks. And it wasn't until I met Andrea that I realized other girls knew the secret, too.

I won't lie—her scrappy way of getting to the top pissed me off, but in the end it only made me better. Not only did I work harder than I'd ever worked in my life, but I measured myself against the best competitor I'd yet encountered in the Navy. I humor myself to think I helped raise her game as well.

Soon the boys came up with a nickname for Andrea and me—the fembots—a reference to the robotic babes from *Austin Powers,* beautiful ditzes who unleashed lethal bullets from machine guns mounted in their nipples. Any time we got together in a class or activity, the boys would say "The fembots are at it again," moving their arms in robotic motions.

While we were graded individually, at VT-86, the class's reputation also rested on how well we performed as a whole. As the saying goes, "You're only as strong as your weakest link." So each person's individual performance had overall importance, and because of this, despite the very real competition, deep down we all pulled for each other and tried to help the entire group along. When you get this far into flight school, it's sad when you see your fellow officer—your friend—not go the distance.

Shortly after Andrea and I were branded as fembots, two guys in our class really highlighted themselves as serious underperformers, and truthfully, their effort simply sucked. They were likable enough and had friends toward the top of our class who could have easily intervened, but instead, our instructors and classmates asked us fembots to try to convince the subpar pair to dig a little deeper.

I'm not really sure why they asked us to do this. Maybe it was because the guys often gave girls the more unpleasant tasks, including things like hosting holiday parties for the children of our classmates and instructors, or cleaning the squadron popcorn machine. They also could have thought that since we were women, we'd likely be gentler in how we dealt with these stragglers. Or, maybe they hoped that if we were in stone-cold fembot mode, we'd machine-gun them down.

We tried both sweetness and tough love, and though I wanted the fembot intervention to work, ultimately, the pair washed out.

"Those guys were self-loading luggage," I heard one of our instructors say about the two after they were gone. Beer in hand, the instructor wrote off their future in the Navy in a short sentence, "They'll be lucky if they wind up as 'tunnel rats' in a P-3."

A P-3 is a large, four-engine prop plane that conducts anti-submarine operations and surveillance for the Navy. P-3s are used for hugely important functions, but by and large, they are big, slow, safe planes and their flights are less intense than jets. My brother is a P-3 pilot. He flies around the world for the Navy on top-secret missions, and he's done everything from busting a million dollars' worth of cocaine down in Central America to surveilling in the South China Sea, getting joined on by aggressive Chinese fighter jets.

So as I got further down the naval aviation pipeline, I was beginning to understand the different reputations and even stigma that came with various aircraft platforms. I knew the course I'd chosen after that first flight with Hardass was going to be a difficult one. Jets was a long subway ride with lots of stops to let people out.

★

Another Jet Girl mentor came into my life at this time. Gringa had gotten her call sign because she herself was Hispanic and she was also married to a Mexican-American aviator. Though she was the Air Force equivalent of an NFO, she'd been one of the few females in the jet community, and one of its pioneers. I suspected that she wasn't a Bonnie, my tormenter at the Academy, but an operator at the core. Like many women in the military, it was more important to Gringa to meet military regulations than to maintain the feminine mystique. Her jewelry was simple and modest, hair and makeup minimal. She didn't have the star quality of J. Lo, but she was still a Jet Girl and an instructor, and I wanted her to like me.

I met her for the first time on an overnight trip to New Mexico. We'd taken two jets up to Taos for training purposes, but since it was the weekend, we got to rent a chateau and ski. These trips, while fun, were not just about arriving at an exotic location, they were a way for our instructors to get to know us and evaluate us. It was on these trips that the real evaluations, at least those that mattered, occurred.

Ashley, my bubbly and beautiful friend from the Academy, had also gone jets. She was three months ahead of me in flight school so once again, I followed her lead from a style and social perspective (she made being a student aviator look so cool); she also taught me all the tricks and tips to succeeding in flight school, as the concepts she'd learned (and mastered—she was crushing flight school) were still fresh in her mind to pass along.

On the first night after we'd arrived in Taos, Ashley and I were stocking the kitchen with the food and booze we'd bought when one of the instructors called out, "Fifteen minutes and we meet downstairs to leave for dinner."

"Fifteen minutes?" Ashley said as we hurried up the stairs. "More like an hour."

Gringa followed us into the room and shut the door. "All right, girls." She looked us dead in the eye. "The guys expect us to be late. This is the way it's going to be your entire career.

They're used to their wives needing extra time, but that's not us. We're gonna beat them back downstairs. Set the tone and they'll owe us the first round at dinner."

Our suite turned into a backstage dressing room during a mid-act costume change. The bags popped open, tossed on the bed. Our ski-bunny chic outfits flew like pigeons across the room. Shower, makeup, a blast of hot air from the hair-dryer, and Gringa, Ashley, and I were back downstairs, drinking Bombay Sapphire martinis, dressed to kill. Or at least to maim. After all, we were in the Navy, so we kept it respectable.

"What took so long?" I said as the guys stumbled down the stairs.

"Yeah, you dudes did some serious primping." Ashley held her glass out for Gringa to clink.

On these trips, it was not uncommon for friends or family who lived nearby to join us during our off time. We all knew that what we were doing was dangerous. You never knew if your next flight or training exercise would be your last, so aviators always prioritized seeing family, even if it meant the family joined the party.

So when my parents, who lived in Colorado, knew I would be relatively close in Taos, they came down for the weekend and met us for dinner at a crowded Mexican cantina. My dad is a pilot himself, and a huge military aviation buff, so midway through our appetizers, he was already grilling Gringa, a seasoned F-15 WSO, on maneuvers. I was trying not to listen to the conversation, but after a raft of questions, she politely interrupted, "You know what—"

I immediately tensed.

"Even after only knowing your daughter a short time, I can see she's going to do great in this community. And after her first fleet tour, she should really think about Topgun."

I nearly choked on my beer. *Me? Topgun?* I sank back into

the booth and hid my smile. I had not even sat in a Super Hornet, much less thought about what I would do in my career four or five years down the road, and yet an instructor—a combat-hardened Jet Girl herself—felt I had the potential for such an incredible challenge.

Maybe she's just being nice. I didn't know. I didn't care, either. True or not, her confidence in me poured fuel onto the fire burning inside of me to reach number one, to fly in F/A-18s and serve my country to the absolute best of my ability.

CHAPTER THIRTEEN

★

March 4–6, 2014; Athens, Greece

Of all the Triple As, I'd probably grown the closest to Taylor. At a time when it felt like no other twenty-something female in the world got me, knowing there was a Taylor around made life easier. Not only was she a fellow WSO in the Blacklions, she, like me, loved shopping, fashion, and making plans, so when we got a little freedom, the two of us went to town.

After two weeks transiting the Atlantic, our first port call was Greece, and Taylor and I booked a cute little hotel room in Athens next door to the apartment the rest of our squadron had rented for the weekend. We weren't snobby, but the boys' apartment didn't have a separate bedroom with an attached bathroom that Taylor and I could lock ourselves in, if need be. Of course no one in our squadron was a threat, but we'd learned that being proactive and staying in our own space protected against unwanted advances or drunken mistakes. There were also added benefits, like not sharing towels with twenty-six other aviators and not finding a "floater" in the toilet. Though Taylor and I never talked about it, I supposed there was some irony that we had to take these precautions with the same group of guys we trusted every day with our lives.

Our first night in port, it was my turn to stand duty. As aviators, we assigned and chose our duties based on an intricate points system, so for our first watch of deployment, Taylor and I chose to return to the boat for a VIP reception being held onboard. As officers on duty, clad in our dress blues, we were there as Navy showpieces for all the Greek heads of state, American dignitaries, attachés, diplomats, politicians, and socialites. The *Bush* was not actually docked in Greece, as the pier itself was too small for a supercarrier. Also, we had a nuclear generator in the belly of our ship, and since many countries do not want a nuke parked on their shore, we had to take tenders to the mainland.

The reception itself was a nice chance for Taylor and me to put on some makeup and brush elbows with some interesting people, the most intriguing of which was a tall, mysterious blonde who called herself Jennifer.

"Excuse me, are you Caroline Johnson?" I felt a tap on my shoulder, just as I was polishing off a cocktail shrimp. I turned to find a well-dressed, perfectly composed woman. Not only was she beautiful, she was glamorous, and she knew my name.

"Yes, ma'am." I offered my hand which was met with a firm handshake.

"Earlier, on a tour of the flight deck, I saw your name on a jet . . . and I just wanted to introduce myself."

Jennifer went on to explain that she was shocked to discover that girls flew jets. We had some champagne and good conversation, and as the food was starting to run out, she said, "You should know, some of the skills you have could be very useful and could serve our country in other ways, if you ever decide to get out of the Navy." She reached into her pocket and pulled out a crisp card. "Keep in touch."

I took it politely. Then glanced down at the embossed CIA logo. Jennifer was a spy.

She read my reaction, turned, and dissolved back into the crowd of VIPs making their way to the tenders.

While I found the encounter amusing, at that point, I was unable to conceive of my life without the Navy in it, so I pocketed the card and hurried to find Taylor.

"If we want to get on a tender, we have to move," I said. We ran back to the Sharktank, threw on civilian clothes, and barely caught the last transport to shore. Overflowing with Sailors, aviators, and party guests, the seating area felt like an escape boat off the *Titanic*. Taylor and I were searching in vain for open seats when I heard a familiar accent. "Don't bother. There are no seats here."

I looked up to see Lorde walking briskly toward us. "Come on, you two. I saved you space in the cabin." He extended his hand, and I couldn't help but smile.

"Caroline," he said as he led us to the cabin, "do tell me. How, at this late hour, do you plan to make it to town?"

"It's not that late," I said stepping over legs and around bodies. "We'd planned to just take the bus to the metro and walk into the bar district."

"Nonsense, ladies, you can't possibly do that. It's much too late. Athens is too dangerous and schlepping it across public transportation will take too long. You can ride with me and my driver."

Given his rank, I knew Lorde had a nicer stateroom on the boat and enjoyed other perks, but up until that night, I hadn't realized what privileges it afforded him.

The first to step off the tender, Lorde whisked us directly into the gleaming white Range Rover awaiting him at the pier.

"After you, ladies," said a tall Greek driver, holding open the door.

We hopped in and hit the bars, with Lorde treating us to rounds of drinks and introducing us to various senior leaders and dignitaries. Having worried a little about Lorde's advances, I'd found myself pleasantly surprised by how respectful he had been that night. I was charmed and grateful. But the night wasn't over.

After a few hours of somewhat formal, stuffy conversation, we managed to drag the high-ranking group to meet up with the rest of our squadron, who'd been letting off steam since earlier that day. I could hear the sounds of carousing emanating from the small Greek bar a full block away. Bartenders were running across the street to buy more beer because our gang was drinking the place dry.

Inside the bar, I slipped to the periphery to take it all in. As pop songs blared from the jukebox and my squadronmates and our bosses toasted with locals, the war in Afghanistan seemed impossibly far away.

"Caroline, what are you thinking about?" A familiar voice drew me from my zoned-out state.

"Oh, hey, Lorde. Thanks for tonight. It's been really . . . great."

"The least I can do. But honestly, what's going on in your head? I'm intrigued."

I shrugged.

"To watch you, I'd guess you're missing a special someone back at home . . ."

"No, not anyone in particular," I said, and he moved in closer, grinning like the Cheshire Cat. I shifted in my seat, trying to create some distance between us but realizing I was in the corner, trapped.

"Are you enjoying all this craziness? Because I was just thinking we could head somewhere more quiet . . . so we could talk. I know a place that's more our style."

Before I could come up with an excuse to get away, I heard someone shout, "Look out!" Beer rained down from somewhere above. One of my friends was trying to pour a beer in another buddy's mouth while standing on a second-story balcony inside the bar. Of course he missed and splattered beer all over the bar and the table near Lorde.

"Good God, man!" Lorde leapt back.

The crowd below cheered for an encore. "Try another!"

This time, the cascading beer splattered mostly on his face and chest. I would not partake but was grateful for the distraction. The waterfall of Pilsner splashed Lorde and allowed me to slip away and grab Taylor.

We dashed out the door before Lorde could catch us, and made our way to our hotel, laughing down the cobblestone streets of Athens just as the sun began to rise.

★

People often ask me, "So, like, what do you do when you're in port? Are you able to do anything fun?" And most are surprised by my answers. First, we do anything you would normally do if you were to visit an (often) exotic city on the water with cash to burn—with the caveat that we also have been holed up for weeks, if not months, so there is an added splurge factor. To answer the second part of the question, it depends on what you consider fun. If you find shopping, getting massages, going to spas, doing sightseeing without making it your job, enjoying decadent lunches and dinners, wine tasting, and dancing with your best friends fun, then yes, you are able to have fun.

Every port call is different. For example, if you could park a carrier in a giant lukewarm puddle of water on the edge of Reno, Nevada, it'd be a bit like pulling up to Bahrain. It gets the job done for people to get their sins out but it's not as glamorous. For those curious, I'd suggest the next time you are in

Cancún to try limiting yourself to low-end all-you-can-eat buffets, warm beer, and staying in hotels with two stars or less and you'll get the vibe. When we pulled into Oman, we were the first aircraft carrier to do so, and it was literally a pier with a bunch of sand, beer tents, and a pizza oven. It was like a giant adult's pizza party in the desert with camel rides.

Turkey was one of my favorites and well worth a trip if you have not been. Our second port call was in Antalya. Our day started off with weird dress restrictions for women which resulted in a lot of female Sailors disembarking dressed in sweatpants and Hello Kitty shirts. The dress code was meant to pay respect to the local culture, not to offend by criminal fashion faux pas, but as soon as we got off and saw the beautiful Turkish women dressed like they were visiting Paris fashion week, we pulled a Superman-style phone-booth change in our hotel rooms and were back on the streets looking like respectful visitors, not a horde of female Michelin Men.

After browsing the local bazaars for scarves and foot mats for the boat, we visited a Turkish hamam, a spa in which all of the girls were laid out on a marble table for body scrubs, followed by a surreal, heavenly ritual called *torba* where the masseuses fill pillowcases with bubbles that they wafted over us before sending us off into the next room for our massages. We healed our bodies, which already were under tremendous strain from taking off and landing our jets on the USS *George H.W. Bush,* but then after all the pampering we transitioned to a decadent ladies' dinner on the stark white cliffs of town which overlooked a dazzling turquoise sea to eat seafood caught that day below us in the Mediterranean.

Later, we went to a nightclub to meet up with the guys, and once again, I was blindsided by Lorde. He materialized at my side like a ninja, immediately hitting me with the hard stuff, skipping the small talk about my previous relationships and straight-up professing his love.

"We'll spend weekends flying around the motherland in jets, our jets."

I'll admit, the life he described (in even more detail than his SERE plan) did sound tempting, and I did find him charming. But the longer I listened, the more I felt physically ill.

"Lorde." I interrupted him midsentence. "Look, maybe if you were not married, and were twenty years younger, and not a superior officer within my chain of command . . ."

I could see his mouth trying to open in defense but I held up my hand and continued. "But by now you should know that's not the way I operate, and that's not the way the US Navy operates."

"But Caroline, I'm not in *your Navy.*" His fingers made quotations in the air. "I'm not bound by your silly rules."

"It's not always about you, Lorde. You have to understand . . . I haven't worked this long and this hard to have an impeccable reputation, just to ruin it in a night. It's not worth it. Even if you do have the most tempting offer in the world. In our Navy, perception is reality."

I looked across the room and found at least two guys in my squadron openly staring. "Like right now, for example. Look at those two dudes over there. By constantly cornering me while we're in port, you're giving them the impression you and I have something going on, which is fatal to my career if you keep it up."

"Then I'm sorry, Caroline," he said, "but that's their problem. And why would you care what they think?"

I held up a hand, feeling woozy. "Maybe it's the warm beer. I'm feeling a little . . . excuse me, please."

I left him at the table and headed to the bathroom to catch

my breath. I took a minute to think about what to do, realizing that if I feigned drunkenness, one of my more sober squadron-mates would walk me back to the hotel, so I pushed through the swinging doors of the ladies' room, intentionally stumbling across the club like I was wasted. As if on cue, Crocket came over. "Dutch, you're wasted. This way." He waved toward the door. "I'll make sure you get home."

Back in my room, I logged onto Skype and tried to call Minotaur, but wasn't surprised when he didn't answer. I Skyped my parents instead, filling them in on the past few weeks and Lorde, unsure how my father would take it.

"Ha," he said, "knew my girl could handle herself with a dodgy old bugger!"

CHAPTER FOURTEEN

★

October 2011; NAS Pensacola, Pensacola, FL

Ultimately the fembot team was split into two separate classes, and Andrea and I were both able to graduate in the number-one slots. The day after my winging, or flight school graduation, the weather turned from sun-washed to almost frigid, clouds rolling in with our hangovers.

The cold didn't bother the Minotaur. "Let's take a walk," he said, driving toward the beach. I was wearing a sundress. Kona trailed along at our heels, occasionally going out to the water and then back.

"Are you shivering?" the Minotaur eyed my shoulders covered in goosebumps.

"Not too bad." I rubbed my arms.

He took off his sweatshirt and helped me pull it over my head. "Here you go."

I inhaled the musk of his favorite hoodie as it warmed me up. "So?" I said. "What's next?"

In less than forty-eight hours, I would have to depart for Virginia Beach to join the legendary VFA-106, the Navy's largest jet squadron. Like me, Minotaur had also received his first choice for the next phase of his career, but he would be on the other side of the country, just north of San Diego.

Dodging me in his classic way, the Minotaur checked his watch, one of those big, clunky diving-style watches that all pilots seem to wear. “Want me to help you get the trailer ready?”

“I’ve got it covered.” I had hired someone to haul the Avion up north because my car couldn’t tow it. He knew this.

“Want to do dinner later? Love to take you to Fish House. Maybe grab drinks at the Bama afterward?” he said, doing his best to avoid the conversation he knew I wanted, but didn’t want to initiate.

Of course I wanted to have dinner with him, but more than that, I wanted clarity. “You know,” I said, thinking of a way to back him into a corner. “My flight school friends are only here one more night. Should I spend the night with them . . . or have dinner with you?”

He thought for a second, toeing the sand. “Definitely dinner,” he said, and my heart fluttered. Then he added, “How ’bout we all go to dinner?”

Before I could respond, he’d peeled off his shirt. “Hang on to this for a second, would ya?”

He handed me his shirt, his phone wrapped in the bundle, then turned and ran out into the waves, his ripped physique disappearing under the water.

Early the following morning, Minotaur came over to help me finish packing my overstuffed Infiniti. He slammed my trunk shut and gave me a hug. “You’ll be all right, kid.”

“I hate when you call me *kid,*” I said, tears running down my cheeks as he pulled me closer. We stood there, hugging, both trying not to show we were crying.

“This isn’t the end,” he said, his lips to my forehead. “You’ll come to California, and I’ll be in Virginia all the time.” He rubbed his calloused thumbs under my eyes, wiping the tears.

"We're aviators. If we can't get across the country, who can?" He stared into my eyes, his face breaking out into that familiar, reckless smile.

I pried myself away and slid into the driver's seat. He kissed me one last time through the open window as I shifted the car into drive and pulled out sharply.

I drove like hell out of the panhandle, singing along to country songs and girl-power ballads, crying my eyes out with my hamster in the passenger seat to keep me company.

★

October 17–28, 2010; Kittery, ME

Survival, evasion, resistance, escape (SERE) school was a two-week course meant to prepare us for what would happen if we were shot down or forced to eject from our jet over enemy territory. It is the most frightening, most realistic, most mentally and physically challenging evolution that anyone goes through, including SEALs and Special Forces operators. But ask anyone who's gone through it, and they'll say it's some of the most valuable training they've ever had.

My first week of SERE school took place in Kittery, Maine. The second week—the fieldwork week—was conducted in northwest Maine at the Redington training facility, basically a giant, military-controlled park. But it feels more like Jurassic Park, in that scattered all over that lush mountainside were things that wanted to kill and eat you.

When you arrive in Kittery, you enter a large auditorium, and the classroom training and lectures commence. But as with all Navy classes, before I began, there was a gauntlet of paperwork, and included in the stack was an oath to sign, stating that there were certain things about my experience at SERE school that I could not disclose. So, yes, all of the information here has been vetted and declassified by the Department of Defense, but there's a lot that happened at SERE school that I won't discuss.

In addition to an oath of secrecy, there was also a two-page form where I basically signed all my rights away. In the Navy, you don't have many rights, but you do have some, and in SERE school, even those go out the window. Things get physical, and you can't retaliate or react defensively, otherwise you fail out of training. With the paperwork complete, the real learning started. Once again, I cannot tell you what occurred during that week, but I can tell you that from the first day, I was emotionally jarred. And that is exactly what the instructors wanted. They wanted us to believe that horrible things would happen if we were captured, because they would. Period. So in that first week, I tried to imagine what a Taliban fighter—the same guy who hunted Navy SEALs in the mountains of Afghanistan—would do if he captured a blond, female aviator from the United States. How would he behave in interrogation? What would he want from me and what would he do to me to get it? How would I resist? Those were the questions I had to ask myself as the course began.

During this classroom period, everything was controlled—when we ate, what we said, when we slept. The stresses were perfectly designed and orchestrated to put everyone on edge. I'm a pretty resilient girl, but at the end of the first day, all I wanted was to talk to my boyfriend, to hear his gruff voice, even if he didn't say much. Minotaur was never a huge talker, but he would always listen. I thought about how much he would love this crap and how he actually could have made a great SERE school instructor. I felt like, for once, he would not only have just listened to me, he might have had some advice. That first week we were in a secured building without our phones, so there were limited times I could even make calls or receive texts. But every time I tried—the tiny-legged jerk ghosted me. I texted, I called, I waited, and then texted again. No response.

Maybe they should bring Minotaur in as part of the mental torture exercise? Here I was, in a professionally crafted mind disaster where I was supposed to be psychologically deconstructed and all I could think about was my boyfriend at a bar in San Diego, sipping on an IPA, sitting next to some surfer girl, watching his phone light up with my name and sending it to voicemail. It wasn't until we were packing up the buses for Redington at the end of the week that my phone buzzed in my pocket.

"Hey, bud. How's SERE school? Havin' fun yet?"

"Fun? . . . fun? . . . let me tell you what is not . . . hello?" I looked down at my phone. No bars. *Damnit.*

"I'm calling you now," he said with a nervous laugh. Here, while I was enduring my toughest trial to date, Minotaur had put me through his own version of SERE, and while I still couldn't shake the image of the surfer chick and the cold beer, my logical side told me to forgive him. Surely he'd just been busy getting settled in California and checking into his new squadron.

The second week of SERE school—the week in the field—was by far the hardest. Before we boarded the bus, we dropped our phones in a basket and headed four hours north to be one with nature for the week. We didn't shower. We barely ate. We drank from antique canteens filled with foul-tasting stream water purified with iodine tablets.

At SERE school, there were no second chances. If I washed out, I could not go back and repeat the course, and therefore I couldn't fly in combat as a naval aviator. It didn't matter that I'd invested two grueling years in flight school learning to fly planes; if I failed a two-week survival course, I couldn't do the job the Navy spent millions of dollars training me to do. It was that simple. If I failed SERE school, I would never fly F/A-18s.

Our first day in the woods, we were given two MREs, or meals ready to eat, and for two days, they let us eat anything and everything out of the ration kits, but after that, we were to survive on water from the creek and anything we could scavenge from nature. We nibbled on tree bark the instructors trained us to find, sampled edible moss, and stared longingly at bare apple trees, picked clean by the previous classes. My group found and killed a plump garter snake. With the instructor's guidance, we hung it up to let it bleed out, but when a Marine slid a knife down the snake's belly to skin it, a handful of baby snakes slithered out, and even the tough Marine jumped back, gagging.

"How does it feel to have performed your first abortion?" a guy behind me joked, but I was far from laughing. I knew what was coming next.

"You killed it," our instructor said. "You eat it."

We finished skinning the snake, chopped it up in a 1950s metal canteen cup with iodine water, and boiled it over the fire we started. There wasn't enough meat to pick off the bones, so the instructor told us to pop the half-inch cubes into our mouths.

"Just chewing it will give you the nutrients you need," he said. "Plus, we never waste a gift we've taken from nature."

Crunching through bones, the snake tasted like salt-free, soggy potato chips, except the slimy, gummy, bland nastiness lingered in my teeth for the rest of the afternoon.

We had competitions for who could light a fire the fastest, and here, I had a slight edge. A couple of summers prior, at Torch Lake with my mom, I was watching a TV survival show called *Man, Woman, Wild*. In the show, a guy and girl are pitted against each other in survival scenarios, and in this episode, the woman revealed that tampons made the best kindling. So, a few days earlier, I'd packed some tampons, even though it wasn't my time of the month, and as the guys scrounged for dry leaves and brush, I sparked up my fire the old-fashioned way—the Tampax way.

Midweek, just as we were wrapping up one of our training sessions, things started to get real. Loud blasts reverberated through the woods, smoke grenades went off, and sirens blared in the distance. Men hiding in woods shot off rifles, and the instructors screamed at us in a thick Russian dialect we hadn't yet encountered in training. We frantically hustled into our groups and took off running up a mountain, trying to escape the simulated plane crash that had just kicked off our "scenario." We had successfully completed the first letter of SERE—we'd survived the crash. Now it was time to do the second step and evade the enemy who was hot on our tail.

The goal is to evade your pursuer for as long as possible because any day spent living off the land, evading, is better than a single night in captivity. If it were real life, we would have continued through the night, trying to escape, but for the sake of safety in training, we got to bed down for the evening. Exhausted and starved, we were all speechless when we saw the miraculous gift waiting for us in the middle of camp—a white, fat, fluffy-tailed rabbit, sitting in a metal cage like he was laying Easter eggs. The instructors didn't need to spell it out for us. If we wanted to eat, we had to kill the bunny, so without hesitation, a couple of students in our group grabbed the rabbit, snapped its neck, and skinned it.

"Who here wants to eat the eyeballs?" our instructor called out. Seeing the horrified looks on all twelve of our faces, he added, "Seriously guys, they're the best part. Eyeballs are full of electrolytes and will help with muscle cramping."

In my small SERE school group, there was only one other

girl besides me. We'll call her Katie. Katie was a Naval Academy graduate a year behind me, a P-3 NFO and a vegetarian who hadn't eaten any meat—not even fish—since high school. But we'd been hiking through the Maine woods with fifty-pound packs on our backs in freezing temperatures. We hadn't eaten anything substantial for days, and her body was about to quit on her; still, I was surprised when she stepped forward.

"I'll do it," she said weakly.

The instructor picked the rabbit's head up by the ears and extended it to her. It still looked perfect, fur intact. I could see the specks of nail polish from Katie's last manicure as she held the rabbit head, fresh blood dripping from the severed white neck.

"How do I do it?" she asked.

"You've got to suck them out," the instructor explained, and a few from our group bent over, dry-heaving at the idea.

She nodded, taking one last look in the little guy's red eyes before putting the rabbit head to her lips, wrapping her mouth around the eye socket and sucking as hard as she could. A red eyeball popped out into her mouth, but it still was attached to the optic nerve, and everybody, I'm talking everybody—tough-guy Marines included—cringed as we watched her sever the optic nerve with her teeth, chew up the eyeball, toss her head back, and swallow.

Oh my God. If I had any food in my stomach, I would have ralphed.

"How was it?" the instructor asked.

"Kinda like an oyster." Katie grinned like a kid on Halloween and dove in for the second eyeball.

The final phase of SERE school is the simulated POW camp, and as with many military training exercises, they saved the worst for last. All night long, babies crying, children screaming for their parents, and psychosis-inducing chants played on repeat. I sat on a wooden box in solitary confinement while strange propaganda blared over speakers throughout the compound. If I had to pee, I had to use the small white bucket in the corner of my cell by the door, but they warned me that if it overflowed, I'd get beaten. If I had to go number two, they took me with a bag over my head to an outhouse that was the foulest place in all of Maine, certainly the foulest place in Redington. They removed the bag long enough for me to hold my breath, run in, and run out. I was forced to drink a canteen cup full of water every hour or so. Of course I was worried about overfilling the coffee can, but they told me if I didn't drink, I'd get beaten. There was no winning.

In the cell, days bled into nights and I lost track of time. Sitting in solitary confinement was lonely and allowed my mind to wander, but it was better than when the door creaked and I was forced to interact with my captors. Each time they entered my cell or forced me with a bag over my head out into the other areas of the prisoner camp, any number of unfortunate surprises or miserable experiences awaited me. Interrogations, beatings, simulated torture exercises, you name it, I did it.

Finally, as our time in the camp was drawing down, I anxiously awaited the door creaking because I thought it meant my time in the cell was complete, but again, things went from bad to worse. I was reunited with the group, all of us shoved onto the frozen dirt, and we were forced to do slave labor, the sharp rocks cutting our hands and knees. We loaded sandbags, patted down roads with our hands, while instructors in character called us "pig-dog Americans," and the females from our group "swines." Over the course, the girls had been pulled aside many times and beaten in front of the men, in hopes of evoking a visceral response from them. But this day, they had an extra special torture for us.

"Latrine," the chief captor said, handing me the filthiest cleaning utensil I'd ever seen. "You, now, clean."

Oh hell no, I thought. *I'd rather be sucker punched again.* But after a few swift corrections for our disobedience, Katie and I complied, crawling on our hands and knees over to this foul, unholy place while our slave drivers verbally mushed us like sled dogs.

In the latrine, I did my best to hold my breath for seconds at a time, but I still couldn't stop the gagging. Cleaning a literal shit hole, I looked up and saw Katie's eyes watering from the wretched gases emanating in our faces.

"This is so much worse than rabbit eyes," she said, nearly crying.

Just when I thought I couldn't handle it anymore, I heard a commotion back at basecamp—smoke bombs, machine guns popping. Our captor continued barking orders until he saw something at the latrine entrance and stopped, suddenly appearing scared. Our rescuer burst into the outhouse, and we ducked into small balls on the floor like we'd been taught. The SEALs had finally come for us.

We followed them into a safe area, not stopping until we heard one of them say, "Congratulations, you have completed SERE school."

Our saviors lined us up and this time, instead of bracing for another flogging, we held our hands out and waited for them to divvy up a huge tray of warm, fudgy brownies.

"Oh my gosh," I said, looking down at my black palms and dirty fingernails and remembering that minutes before I had just been scrubbing the latrines from hell. "Guess there's no sanitizer," I asked, knowing the answer.

So I did what I had to do, shoving aside all of the clean-freak, OCD tendencies of my debutante past, and with my putrid hands, I ate the brownie that, as far as I'm concerned, should have been immortalized in the *Guinness Book of World Records* as the

World's Most Delectable Treat. In fact, I think that brownie should be somewhere in brownie heaven, glowing, sainted.

★

In the military, when you return to civilization after SERE school or another one of these rites of passage, you often find yourself gravitating, almost clinging to the people who were with you. After leaving SERE school we all went to Boston and met up to drink.

Even though our group hailed from different operator communities—aviators, spec ops, Marines—our group pulled in close together, moving like a school of minnows through the streets of Boston. There was an understanding among us, even though we were smiling, drinking a beer in a packed-out bar, and dancing, we all knew that we were not really there. We did our best to enjoy ourselves, but our heads were still in a bubble, hours away. In a simulated POW camp. Sucking the eyeballs out of the rabbit.

Civilians just don't understand this. So that night when a couple of cute boys moseyed over, there was no way I could deal with it, no matter how charming they seemed. I couldn't stomach their bullshit small talk. I didn't care about how much they would make as management consultants, or how exclusive and cool their Harvard B-school section was.

One of my Marine SERE partners, watching from across the room, came over and tapped the guy on the shoulder. "Hey, dude, think she's done talking now."

"Yeah," I added. "I'm just gonna chill and finish up my beer."

The pair of yuppies looked down at me spinning on the barstool and then sized up my friend, trying to figure out if he was someone to mess with or if they should stand down. "Whatever," he said turning away from us.

"Thanks for the rescue," I said when Mr. Sport Coat had migrated down the bar.

My friend nodded and we sat there together in silence, the noise still pulsing all around us.

★

Upon returning home to Virginia Beach, I had ten days off before my next training cycle. A longtime family friend from the Naval Academy was getting pinned as a Navy SEAL, so I wanted to be there to support him, but if I was honest, there was another reason. The ceremony was on the island of Coronado, just over the bridge from San Diego, where a certain skinny-legged hunk would be waiting.

After his stellar boyfriend performance while I was stranded in Maine, you might wonder why I wanted to see Minotaur at all. But in truth, he was a tough guy to hate or even be angry with. I liked that he was fundamentally a quiet and reserved person. Returning from Iraq, he'd survived IEDs and ambushes and had lost many of his Marine brothers who fought with him in battle, so naturally he was a little raw. Going into the relationship, I knew his shortcomings and complications and past. I went into things eyes wide open, and I went out to California eyes wide open as well. I hoped things would change, but I had no expectation they actually would.

We spent a week at the beach encased in the golden light of Coronado—swimming, lying in the sun, and going back to our hotel room. We never fought, because in a fight, you have two people attacking each other, but when it came to emotions, the Minotaur never attacked, he just withdrew.

Perhaps it was the courage I'd garnered from having such a nice week together that prompted me to open my big mouth that afternoon.

Already embracing the aviator style at a young age. Pilots and WSOs are known for sporting stylish sunglasses, which protect our most valuable asset: our eyes.

Skiing with Dad and Craig. Even before I could walk, I remember extreme ski adventures with my family. As expert skiers, my parents always challenged us but also ensured we had fun along the way, which fostered my adventurous, daring spirit.

My family is very important to me and we always took advantage of any days I had off from the Navy to spend time together.

EXCEPT WHERE NOTED, ALL PHOTOS COURTESY OF THE AUTHOR.

Showing off my plebe dorm room at the Naval Academy. The Annapolis heat and humidity weren't kind to my plebe haircut.

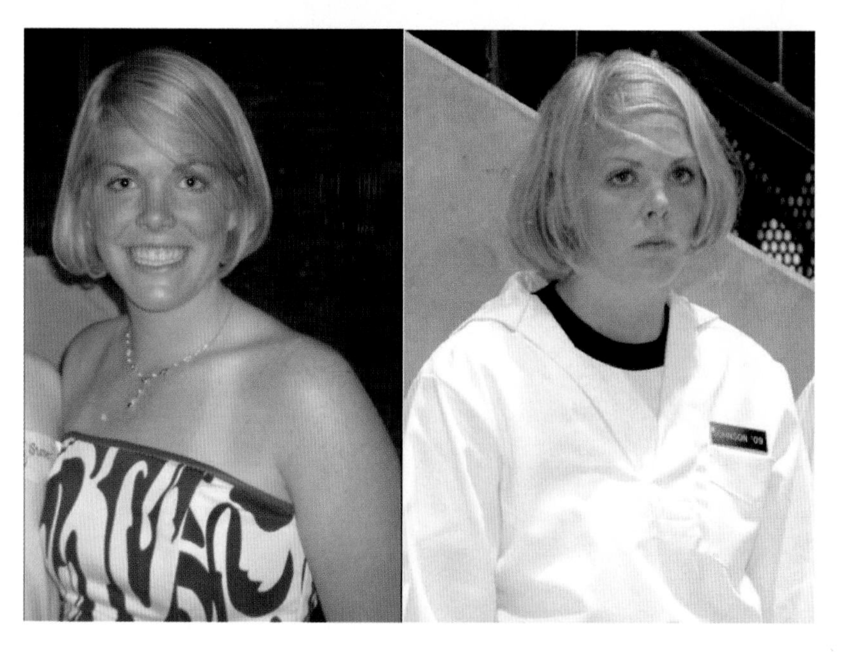

What a difference forty-eight hours makes. Before: enjoying a fun debutante cocktail party. After: my first day at the United States Naval Academy after being indoctrinated as a plebe.

Ali and I were best of friends through the Academy and into flight school. Here we are with our swords getting ready for noon meal formation at USNA.

All dressed up with nowhere to go but to sea. My mom surprised us with festive hats and custom T-shirts to wish the Sharktank well on our nine-month deployment.

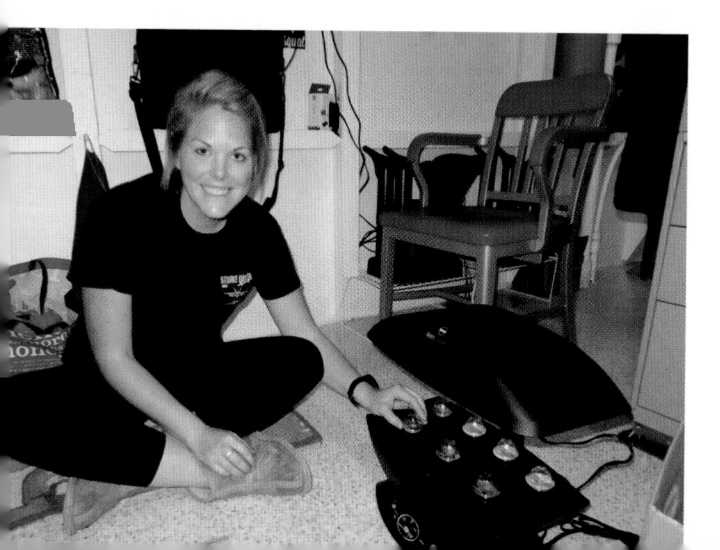

Tending to our herb crops in the Sharktank. I grew fresh herbs in a hydroponic herb garden to brighten some of the ship's unfortunate meals during my nine months at sea.

My Naval Academy friends embracing the challenges of plebe year.

Enjoying time with my mom and Grammy, two of my most powerful female role models, after my winging ceremony.

I received amazing support via care packages, letters, and messages from the home front and always made sure I shared.

Learning how to properly maneuver the models is an important part of briefing basic fighter maneuvers (BFM), or dogfighting as most people call it. I spent many days and hours studying at my desk in the Sharktank.

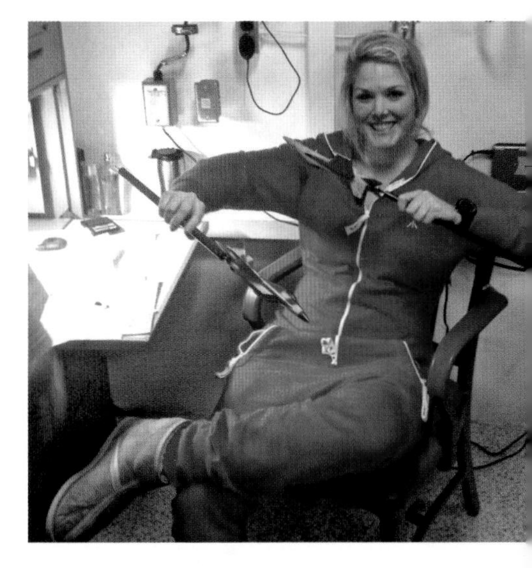

Ecstatic to have completed flight training and earned my wings of gold in October 2011.

At the Minotaur's soft patching ceremony. We winged on the same day, at the same time, at separate airfields—which meant we couldn't attend each other's official ceremonies, so the more informal, soft patching had to do.

Also halfway through my plebe year I got to celebrate another milestone. I was selected as a debutante and was honored at my debutante ball. Getting back to my girly roots after six months of having my identity stripped of me in the Navy was a welcome relief.

Monkeying around in the Sharktank. Living in close quarters with six women, we had to get creative. This hammock was well utilized during our training and deployment on the boat.

The Caroline/Carolyn duo sporting our fleece vests with the Rock of Gibraltar in the background.

Showing off my flight gear in the paraloft where we got dressed as part of our pre-flight ritual.

The Sharktank in our stateroom. We were very fortunate to have such a spacious, well-decorated home to escape the industrial environment of the aircraft carrier. The Jet Girls are in tan; the E-2 bears are in green.

I'll be your wingman anytime! In the fighter community we rarely fly alone. We always try to be good wingmen and support the team we're flying with.

Three of my avionics technicians on the first day of Iraq combat operations in 2014. These men and women were so talented and hardworking, and most were no older than twenty-five.

Sorbet sunset. Flying over Western Afghanistan we caught an incredibly colorful sunset above the undercast layer of clouds.

After nine long months we're finally home! I was lucky to have family, friends, and my Jet Girls come out to the flight line to welcome me back to Virginia Beach, VA, on a blustery November morning.

Enjoying a much-deserved brunch and weekend of R&R in Dubai after seventy-two days at sea fighting ISIS. Myself, Carolyn, and the rest of our squadrons enjoyed our port calls and experiences immensely.

Light reading in flight.

Flying with my mentor and good friend Deerlick on an "unmanned aerial vehicle," or UAV, flight. Deerlick was an instructor of mine in flight school and showed me the ropes in the F/A-18 community.

Celebrating another safe and successful flight and another day down on deployment. Our plane captains (brown shirts) and senior maintainers (white shirts) were essential members of the team who got our jets flying and kept them in the air every day.

Getting dressed and ready for our combat flight. Flight gear: check. 9mm pistol: check. Brown bag lunch: check. Morale-boosting glasses: check. Smiles: check.

Flying with my dad. My passion for aviation runs in the family!

Jumping for joy that we finished our transatlantic crossing and were finally in the Mediterranean Sea on our way to combat in the Middle East.

My first and last flight in an F-16. The Navy uses a fleet of F-16s and F-5s, among other jets, to simulate our adversaries and test aviators' visual identification skills. Fallon, NV.

Dudeboat and me after manning up as standby participants in case any of the primary players went down during our first airstrikes into Syria.

Fresh out of the jet as a standby participant for the first strikes on Syria. The US news stations were covering our strikes before the jets even returned from their mission.

Me and Grammy. My Grammy has always been a role model and inspiration for me. I was so proud to be able to fly in to visit and show off my jet to her.

Launching off the USS *George H.W. Bush* into combat.

When a hurricane rolls in and you can't do the flyover for the Navy vs. Air Force football game, you dress up in your best fan gear and enjoy the game with your dad.

My best day at the Naval Academy: graduation and commissioning on May 22, 2009, with President Obama. *Photo credit: © Lenny Lind 2009*

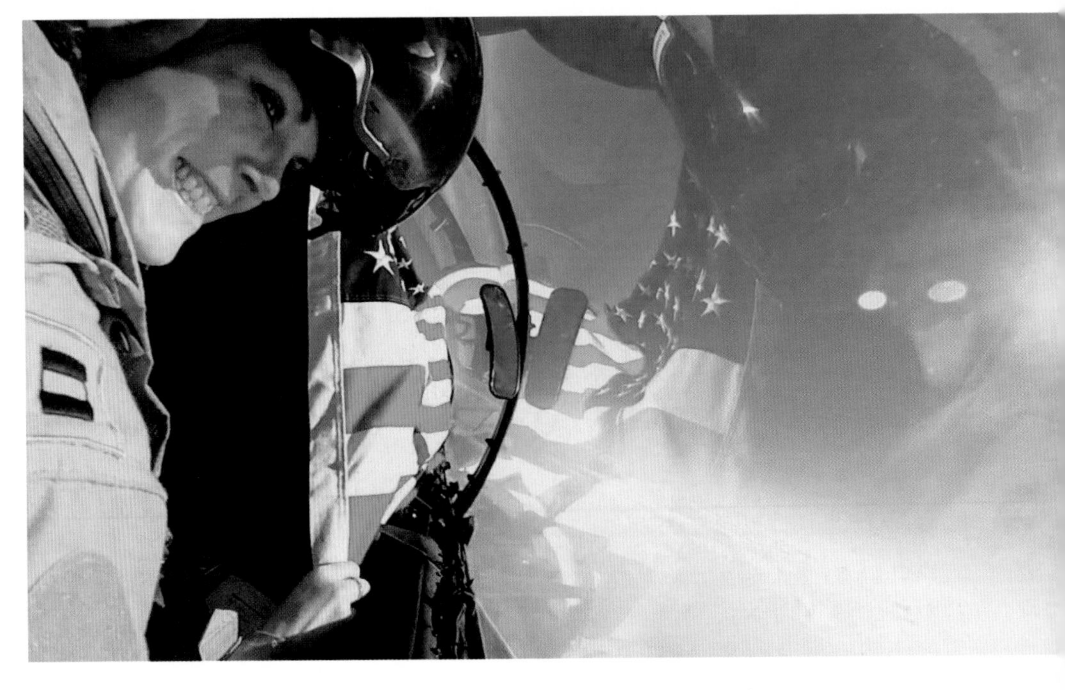

These colors don't run. Flying the stars and stripes above Iraq.

"So, Christmas," I said, slowly folding my clothes. "You have any plans?"

"Probably just gonna hang around here."

"Really," I said, acting like this was news to me. "Well, I'm gonna be visiting Grammy in Arizona."

He flipped the page in his magazine. "Yeah, she's awesome."

"You know it's only four hours away, right?" I wasn't about to ask him to come see me. I didn't need to. The Minotaur always knew what I was thinking.

He looked up at me but said nothing.

"Fuckin' A!" I slammed my suitcase shut. "You know what. I don't need you to drive me to the airport."

CHAPTER FIFTEEN

★

March 2014; Embarked USS *George H.W. Bush,* middle of the Atlantic Ocean

Encounters with Lorde aside, there can be a lot of sexual tension in a jet. For all the seriousness of flying, there's something seductive about it, and perhaps the dangers and the difficulties make this even more so.

The physical act of flying is often reserved for those in their prime. The Navy, and the military in general, is run by junior officers—men and women between twenty-two and thirty—most of whom are in excellent shape. There are weight- and body-measurement requirements that must be met, and if the Navy's standards weren't high enough, most JOs (especially the ones in high-performing functions, like in the aviation community) have the kinds of personality types that obsess about their own workout goals, almost as much as they would jobs and studies.

Take Sully for example. Sully was a pilot who I came to respect and admire. Sully looked like a superhero, almost cartoonishly handsome. He was an all-American farm boy and when he took off his helmet it was hard to take him seriously because he looked like an actor playing a pilot. And he was kind, funny, and deeply respectful of everyone—women and men. He was also one of the best pilots on planes of every variety—jets, turboprops, aerobatic planes, commercial airliners—he had flown them all. He was married, with a young family and a beautiful wife, which almost made him more attractive. He was a safe guy to want to fly and flirt with. All the girls I knew had slight crushes on him. Even I felt a flutter in my stomach every time I was with him in the jet.

There is the flip side to the coin. In aviation, we operate sitting down in cockpits, and while the work does take a physical toll, it's largely cerebral. Unlike the other branches, our missions often require mental dexterity rather than brute strength and physical endurance. Also, as counterintuitive as it sounds, aviators perform better during the most physically demanding regimes of flight if they're slightly out of shape. For example, when I was a few pounds over my skinny weight, I was better at pulling Gs, or resisting the gravitational forces in the jet, and when I was in my best running shape, I was more prone to blacking out. This wasn't unique to me; most of the Jet Girls found similar benefits to being slightly above their ideal weight. When you have higher blood pressure and a higher body-fat percentage, it is easier to resist G-LOC.

But even though extreme maneuvers were easier to withstand with a few extra pounds, some aviators took this idea to extremes. They'd long stopped playing sports after the Academy and discovered drinking and late-night binges. Ironically, these guys were often the ones with trophy wives back home, so while the former sorority girls were maintaining their abs and spray tans, their husbands wound up bloated, pasty, and thanks to the constant helmet-wearing, prematurely bald. Yes, there were jet pilots who looked like Tom Cruise and some who looked like Goose. But there were some who looked like Danny McBride—top of the head shaved, a mullet in the back, chrome dome, Doritos in their teeth, and a deployment mustache. These guys who let themselves go to pot were more often total assholes, which made them easy to resist. In any case, the Dorito-eaters were the exceptions.

Not only are the majority of officers young and in shape, aviators—and this goes for both the men and the women—are very smart and also somewhat aggressive. We spend enormous amounts of time together, risking our lives in tight quarters. But there's another aspect that speaks even more deeply to an aviator's reputation for seduction, and that is our voices. Even the tersest pilot or WSO is a trained and skilled communicator. We choose our words carefully. We listen intently. We give our wingmen and allies full attention and respect in the plane, and when there is danger and uncertainty, often the voice speaking in your ear is the one that calms you down and keeps you safe. It brings you back to the mission or puts you on alert.

Just like in real life, I would argue that in the air, girls are better communicators. Better with pilots in the cockpit, better with the ground forces, better with other planes in the air, better with wingmen, and better with support aircraft. We tend to speak more clearly, giving accurate and precise descriptions even when pulling Gs, straining to make our radio calls.

It doesn't matter where you are on the globe, it's extremely rare to hear a female's voice coming from a fighter jet. When a woman does call on the radio, she automatically has an audience. Often when I checked in with our troops on the ground in the Middle East, I'd notice an echo every time I transmitted. I figured out the echoing was the result of our guys on the ground turning up their speakers to broadcast my radio comms to enemy forces. This tactic tormented our enemies who, generally speaking, discriminated against women, forcing them to hear a woman in a fighter jet overhead—a jet that could vaporize them in an instant. And for many of our troops who hadn't heard a woman's voice in weeks, the Jet Girl cadence was a strange and welcome comfort.

Which brings up another point: Sometimes it's hard to avoid sexual innuendo in a jet. So much of what we do in flight is erotic in nature. Take refueling, for example. We literally stick a nozzle probe into a basket, which, back in Vietnam, tanker crews decorated like a woman, spread-legged. Or the act of targeting the enemy and dropping a bomb involves a lot of weaponeering and manipulating switches and buttons in the cockpit, one of which is nicknamed the nipple switch.

The guys weren't the only ones who'd joke from time to time that flying an F/A-18 can be sexy. Combine the sheer power of the vibrating jet seat, with the tension from straps crossing your waist and legs, with the pilot's cool, commanding voice, and I don't know a woman who, if they're being honest, can deny that there's an element of arousal. After an intense flight and night landing where you slam into the deck with more force than most planes can endure, somehow managing to pin it perfectly, after the helmets come off and you're back in the paraloft, it's almost impossible not to feel something nagging at you.

Don't get me wrong, while flying in the jet with another guy can be sexy, it can be extremely unsexy, humiliating, or downright gross. It's not uncommon for a pilot, particularly one of the Dorito-eaters I described earlier, to have horrendous gas in flight. You see, as we climb up in the atmosphere, like in a civilian airliner, the jet pressurizes, and the increased air pressure makes the gases within your body and intestines expand, so they have to go somewhere. The best thing to do is let go, otherwise, you'll have major issues. (Yes, I'm telling you to light it up on your next cross-country flight—you'll feel better, and it's good for your body.) But in the cockpit, my squadronmates always seemed to follow this gastrointestinal advice, turning the tiny bubble we shared into their own personal Dutch ovens.

That fact, combined with my sensitivity to smells, won me the call sign Dutch.

Many of the guys in the F/A-18 community seemed like overgrown fifteen-year-olds with their constant discourse about pooping and farting. Initially, I chalked this up to social immaturity, but as time went on, especially on deployment, I realized that it was more than that. They talked about these things so that when nature called in flight, they were comfortable enough to deal with the outcome. When you think about combining nasty boat food, pressure changes, copious amounts of coffee, chewing tobacco, and sitting for nine or ten hours without a toilet, the result is bound to be disconcerting for most of the men. In a fighter jet, not surprisingly, there weren't many good outcomes if you felt that tingle, so all aviators carried Imodium, in case the urge struck. There were quite a few instances, however, when Imodium didn't fix the problem, and these moments served to further solidify the male aviators' biggest fears. No, it wasn't getting shot down or crashing into mountains, or even encountering rogue Iranian fighters with live missiles. What kept us up at night was something much worse than that—having to go number two in flight.

Much like Lorde's survival plan, each male in our squadron had intricate poop protocol, in case the unthinkable happened. And those who didn't have a poop plan, well, it wasn't long before their call signs became epic: POMPOM (Pooped On Myself, Proceeded On Mission), POAT (Pooped Over Afghanistan Twice, which then became Thrice), and perhaps best of all, Pooudini. In this instance, a single-seat pilot, while in combat, managed to unstrap, strip off his gear and flight suit, number-two into his helmet bag, and then somehow dress himself again and finish the flight. Thankfully, I had pretty good sphincter control, so instead of stressing, I kept some Imodium in my bag and forgot about it. But peeing was a different concern.

A surprising truth that I don't often tell is that one of my most memorable flights on deployment didn't involve a battle with a foreign enemy, but one against my bladder. When operating in combat over countries like Afghanistan, we launched from our carrier parked in the North Indian Ocean, flew an hour and a half or two north to get in-country, operated for three or four hours, and then flew a few hours back home. Staying airborne for this long required a couple of midair refuels, but unlike your typical road trip in a car, we couldn't make pit stops for snacks or bathroom breaks. Bringing food in the jet was relatively easy, but tackling the second problem was something else entirely, and this is one area where I concede that men had the definite advantage over Jet Girls. Thanks to convenient plumbing and a handy container called a piddle pack, the guys were able to relieve themselves with relative ease. Piddle packs—large, two-liter bags with resealable funneled openings at the top—were filled with a crystalline powder that turned the pee into gelatin. For women, it was more complicated, and the first person who really wanted to talk to me about it was my awkward skipper—before deployment even began—thinking he was helping out.

I was in the Ready Room after one of our meetings when the skipper, who'd just been promoted from XO, pulled me aside. "Hey, Dutch, got a sec?"

I turned to see Coma. He had the nervous look and awkward tone of someone who wants to seem respectful of your privacy while at the same time is trying to pry. A state which, I have to say, was not unusual for Coma.

"What's up, sir?"

"You know," he said. "We're going to be flying in combat, and the flights are really long . . ."

"Yes." I nodded.

"And you've got to make sure you stay hydrated."

"Yes, sir, I know. What are you getting at?" I asked, feeling a bit like my high school gym teacher was taking me aside to talk about the proper usage of contraception.

"You can't just divert to go to the bathroom. Before we fly in-country, you need to make sure you are stocked up on diapers."

"Excuse me? I'm not planning on bringing an infant with me on our long flights, so . . . don't think those will be necessary."

"No, Dutch, adult diapers. Depends or whatever they call them."

"I appreciate your concern, but let me ask you, do you piss on yourself and then sit in it for hours at a time?"

He laughed, but when he didn't answer, I pressed. "Well, do you?"

"Of course not. That's disgusting."

"Good, because I'm not, either. I talked to some other Jet Girl instructors, and they told me about the AMXD." The AMXD was a device created so women could pee in the cockpit.

"Dutch, that thing doesn't work. I'm telling you, you need to wear a diaper."

"Sir, no, the old one doesn't work well," I explained, proving I'd done my research. "But there's a new one. Taylor and I tried to ask the PRs for it, but they said it's too expensive. Maybe you can intervene? Tell them they're authorized to order it so that we can get it in time to fly in combat?"

Coma went silent, an intelligent man who wasn't often at a loss for words, debating his next move. "Uh, yeah, I'll uh . . . see what I can, uh, do."

Upon seeing my satisfied smile he added, "But make sure you bring Depends as backups."

From that conversation forward, I knew Taylor and I wouldn't be getting the new system, so we went to the parachute riggers and picked up old AMXD kits. Even though the old kits cost about $3,000 apiece, for some reason they were okay with ordering multiples of those instead of a couple of the newer, more expensive, advanced AMXDs. Between me and Taylor, we needed at least five kits to piece together two complete sets for ourselves, because the original AMXDs were so old and had sat on the shelf for so long that, in any given box, half of the parts and pieces were defective—rubber tubes dry rotted or lithium batteries leaked all over the place. Once we had the AMXDs rigged up, we had to figure out how to use them. Without a VCR to play the antiquated tutorial, we had to rely on trial and error.

We dumped out all the tubes, pumps, and batteries, studying them like an advanced Lego kit for fighter pilots, ages eighteen and up.

"Think I'd have more luck putting together a bookshelf from IKEA." Taylor snorted at the sea of parts we'd spread on the carpet.

The main brains of the system were contained in a little black box unit that was similar in size to a videocassette. Two tubes attached to either side of that. One tube connected to a pee collection bag, and the other to the collection apparatus. The collection apparatus was just a fancy name for the most unwieldy feminine pad you've ever seen. Not only could this thing fit Big Bertha, but as an added bonus, there was a foot-long tube attached to the pad via a hard connector piece that was supposed to sit at the lowest point in your underwear. The first time I tried it on, I stuck the thing in my underpants as directed, but it was so huge, it poked out of the top and bottom of my Victoria's Secret panties. Still committed to the trial run, I quickly zipped up my flight suit, got the shoulders of my green onesie on like normal, but when I went to zip it up, I found the tube hanging over the front opening, caught up in the metal zipper. With no other choice, I flopped the hose with its hard plastic tip into my flight suit, positioning it so it hung down the right side of my leg. Now I knew how my well-endowed counterparts felt every time they got dressed in the morning. Once I finally shuffled out of the stall, I turned around and checked my ass out in the mirror to see how bad it was.

Now, flight suits are thin, showing absolutely everything going on underneath them. Even though we didn't wear them tight like the Blue Angels, you could still see clearly enough to make the mind wander. Truth be told, I almost always wore lace thongs or seamless boy shorts because they laid flat on my tush and the boys couldn't detect any panty lines. Also, my underpants always matched my sports bra, which also always matched my socks, because as a girly girl, if I had to wear a baggy, olive green potato sack every day, I had to keep it feminine underneath.

Looking at my reflection that first morning, the crunchy plastic pad gave me a distinct bubble butt. Luckily, I was going from the bathroom straight to the PR shop to put my other gear on, and I had my helmet bag with me. Throwing the bag over my shoulder and holding it so that it covered my backside, I weaved through the crowded Ready Room. At the time, I had no idea why the pad was so big, but I'd find out one beautiful April day in the skies over Afghanistan.

Crocket and I had been supporting troops on the ground and were called off of our tasking and onto a TIC, or troops in contact. That meant there were ground forces nearby that were actively engaged with the enemy. That day, the guys on the ground happened to be Navy SEALs, and even though their call for help came at the end of our VUL time, or shift in-country, the Taliban, high on a mountain ridge, had pinned the SEALs down in a firefight and we needed to support them. We called to request authorization from the boat to remain on station to help, and once approved, we were circling overhead, searching for the enemy when we found another group of bad guys on the ground. The SEALs had neutralized the Taliban on the ridgeline, but now a group was pestering them from a wadi, the native word we used for a riverbed. We located the SEALs' target, and while we all wished we could have lit the bad guys up with our 20mm cannons, because of our rules of engagement, we couldn't actually fire without first attempting to de-escalate the situation with a show of force.

We accelerated our jets to more than five hundred miles per hour and descended to about two hundred feet, which is extremely low. At that altitude, we executed a sneak pass, meaning we ingressed the target area so low and fast that they didn't know we were coming until we tore across the sky, kicking out flares in full afterburner in a show of force so mighty and loud that it dispersed the toughest of fighters. The sheer sound and sight of an F/A-18 barreling overhead was enough to knock the enemy to the ground, literally tumbling over like bowling pins when our jet roared past.

If the Taliban did not disperse with a show of force, the next step was to use lethal force—finally matching guns for guns. On this day, thankfully, we didn't have to. After our exhilarating show of force, we saw the weasley fighters disappear like cockroaches when the lights snapped on. The SEALs said the ground force commander's intent was met, the enemy was neutralized, so they thanked us over the radio and sent us on our way.

This had been one of my longest and most exciting days so far in combat, with an incredibly climactic end to the flight. As we headed to our last tanker to refuel for our transit back to the boat, I knew we'd be flying for another two hours and there was absolutely no way I was going to make it back without wetting my pants.

"Dude, Crocket," I said. "I'm so sorry, but I've got to go pee."

"Oh God, Caroline." He sighed into the ICS. "It's your first time, isn't it?"

"Yup. And I'll bet you remember your first time," I joked, trying to divert his mind to other things.

"Yep, and yep," he said, laughing. "I'll let you do your thing back there. Just let me move these." He reached up, shifting the rearview mirrors we used to maintain eye contact between the front and back cockpits. Thank God I was in the plane with Crocket that day. Crocket was my favorite pilot. Smooth, ultra-talented in the cockpit, and outrageously funny. He helped me get through long days of combat. He didn't care what people thought, so he was never vying to fit in.

"As long as you're good at your job and you're not a douche, you're okay by me," Crocket told me once. If I had to pee in front of any dude in my squadron, thank God it was Crocket.

Once Crocket moved his mirrors, I disarmed my seat so I wouldn't inadvertently eject the two of us in case my AMXD contraption got caught in the ejection handle. Next, I unstrapped the upper four connectors from my chest and waist, and unzipped my survival vest. My combat survival gear was so bulky with my 9mm pistol and extra ammo strapped to it that without unzipping the vest, I couldn't see to unzip my flight suit from the bottom up. I dug in my helmet bag, retrieving my pee bag that contained my AMXD, wet wipes, and hand sanitizer. I got the contraption out, and pieced it together, as carefully as one might handle the components of a bomb. With my flight suit unzipped, I tugged on the hose that was now stuck in between my G-suit (the green, ass-less chaps that help us pull more Gs), flight suit, and my leg. Once I got that out, I attached it to the other side of the AMXD, and with the contraption in place, I spent the next five minutes willing myself to pee.

"Come on, come on," I whispered. I knew the battery wouldn't last long, so like the other Jet Girls recommended, I waited until right when I started to pee to flip the switch, but instead of suction, I felt air flowing in the opposite direction. Checking to make sure the switch was set correctly, I realized what was happening. Apparently the purpose of the granny panty–sized pad was so that the thing could create a seal all the way from the front of my pelvis around to the back, up over my butt crack, held in place by an inflated outer rim that filled up like an inner tube. Finally, after about twenty seconds, the seal was inflated and the pump finally started sucking, but just as it was starting to work enough for me to let go and pee at full force, the power died. I wasn't wearing my mask, otherwise, Crocket would have heard a string of expletives like never before. I stopped midstream, fumbling for the extra batteries in my pee bag.

Since I had the antiquated, original AMXD, the long-life lithium ion batteries were all dead, so I had to use battery packs that took six triple-A batteries that would be swapped out multiple times each flight if we wanted to go. I slid the battery pack off the black box and tried to carefully open the thumbnail-sized lid while still holding my pee midstream, but the darn thing was jammed. I jarred the lid loose, but it was spring-loaded and shot across the cockpit. I lunged, trying to catch the small plastic piece as it fell, but as I did so, we hit a patch of turbulence and three of the fresh batteries fell onto the floor of the cockpit, well out of reach.

I put my mask up to my face. "Oh. My. God."

"What's going on back there, Dutch?"

It was one of the very few times in my life that I wished I were a dude. "Crocket, the batteries died . . . midstream. I was trying to change them and the lid went flying and now three of my new batteries are on the floor."

Crocket knew the urgency of the situation. Peeing was one thing, but loose objects, foreign object debris (FOD), were forbidden in the cockpit because they could get jammed in the flight controls and cause the plane to crash.

"All right, Dutch, on your count, hang on. I'm gonna go inverted."

Once I had all of my things secured in my helmet bag and my lap belts back on, I gave the countdown. "All right, three, two, one!"

Rolling smoothly over our right wing, Crocket executed a perfect aileron roll. With the jet flipped completely upside down, he unloaded to make all of the FOD fall into the canopy so we could grab it. This wasn't my first time for this maneuver, but it was the first time I'd done it without my straps tightened down. I drifted out of my seat, suspended like an insect in a spiderweb, and came to rest in the canopy with my head and body squished into the glass bubble, the batteries rolling around just in front of me. As Crocket and I both reached to grab them, I felt the urine rush out of the AMXD tubing and trickle up my body. Just as I saw little droplets appearing on the canopy, below me, Crocket and I snatched all the parts and within fifteen or twenty seconds, he rolled us upright again.

When I reassembled the device, it wouldn't switch back on, and by this point, I was nearly doubled over in pain from my desperate bladder. I pulled my knife out of my survival vest and cut a tiny little hole in the seam of my flight suit. Channeling my inner yogi, I contorted myself into a pseudo standing position to use my backup funnel and a piddle pack secured in the ejection handle to relieve myself. I had to pee as slowly as I could so that the funnel didn't overfill and spill onto any of the electronics in the cockpit.

Crossing the Afghanistan-Pakistan border and mortified by the entire affair, I started to worry that my flight suit wouldn't be dry by the time we got back to the boat.

I'm going to get some horrible new call sign like Dirty Pee Pants Dutch, I thought and spent the rest of the flight home unstrapped, hovering above my ejection seat, fanning my damp ass. Thank goodness my flight suit and the rest of the cockpit dried by the time we landed on the carrier, but that was the first, last, and only time I would even attempt to pee in a plane.

Back in the Sharktank, I told Taylor what happened and she wiped away tears from laughing.

"But what will you do if it happens again?" she asked.

"Easy," I said. "Dehydration is my go-to from here on out, so in ten years, I was hoping you'd donate me a kidney."

In-Flight Dining for Combat

Self-packed lunch with one small pouch of dry packed green olives (sent from home), one bag of Cheez-Its (sent from home), one English muffin sandwich with ham, cheese, mayo, mustard (Navy provided—eaten warm as it was melted by summer heat before takeoff creating grilled-cheese effect), two packs of Scooby Doo fruit snacks (sent from home—one for me and one for my pilot), and a liter and a half of water. **no Skittles, M&M's, trail mix, or any other small, hard items that could be spilled and potentially FOD the cockpit**

★

Combat Flight Gear & Accessories

Wardrobe

Survival gear on my body: matching socks, underwear, sports bra, DriFire fire resistant T-shirt, NOMEX fire-retardant flight suit, anti-gravity suit, ejection seat torso harness, survival vest, salt water–activated life preserver unit, fire-resistant flight gloves, Red Wing aviator boots, oxygen mask, joint helmet mounted cueing system—$200,000 helmet with a built-in targeting system.

Survival gear in my survival vest: oxygen pressure regulator, two radios and GPS beacon, flashlight with red, green, white, and infrared lenses, flares, knife, all-weather mittens, signaling mirror, industrial D-ring to clip into survival hook in case the helicopter had to pick us up after ejecting, Beretta M9, an American flag bandanna (to wipe sweat out of my eyes), small eight-ounce bottle of water for survival.

Survival gear in my anti-G suit pockets: two ammunition clips with fifteen rounds each to use if forced to eject in enemy territory, one liter of water for emergency, KIND bar (sent from home).

Accessories

Survival gear in my helmet bag: night vision goggles and bracket (if landing at night), spare clear visor (if flying at night), blood chit, two CLIF Bars, AMXD urination device (completely useless), backup batteries for AMXD, backup funnel and piddle pack for urinating (pee bags), hand wipes, brown-bag lunch, NATOPS pocket checklist, two error-code books to troubleshoot the jet, kneeboard with normal checklists and tactical checklists with a pen tied to it, charts and tactical mission products, Velcro leg straps to hold charts and tactical products to leg, four magazines—two educational and two smut magazines (*Architectural Digest, Travel + Leisure, People, Glamour*), divert airfield packs, emergency overnight kit (toothbrush, credit card, international cell phone).

CHAPTER SIXTEEN

★

December 2011; Naval Air Station Oceana, Virginia Beach, VA

It was early in the morning. Low forties. Clear blue skies with not a cloud in sight; a "bluebird day." Perfect for flying. I'd finally finished F/A-18 ground school, and I was about to get a Christmas present of sorts—my first flight in a Super Hornet.

The skipper barreled into the Ready Room like a dad about to take his kid on their first roller coaster. "It's time!" he said, clearly excited for me, but also ready to strap into the fun seat.

We cruised through the brief, the preflight, all our checks, and taxied out onto the runway. I did everything, repeated everything exactly as I had done for two months in the simulator, but this time, it was real.

We were cleared for takeoff. The nose of the plane pointed directly down the runway. "All right, Caroline, you all set?" the skipper asked.

"Yes, sir!"

"Hold on then, here we go!"

The plane rocketed forward, and immediately I felt a stack of bricks drop on my chest. *Breathe,* I coached myself, my body becoming one with the seat. *Slow and steady breaths.* My vision blurred, and before I knew it, we were airborne, banking to a new heading, tearing across the sky. I felt a bump and a thump, and a brief moment of panic before I realized it was the landing gear retracting.

Screaming at the speed of sound in a gunmetal-gray rocket ship, I clung to what I liked to call the holy-shit handles, physically holding on for dear life as my mind raced to keep up with the jet. Since it was winter and the skipper didn't want to put me through the pain of wearing a dry suit for my first flight, he took me to the Navy Dare training complex, a bombing range, perfect for showcasing what our sleek, gray stallion could do. Navy Dare is on a small peninsula that sits near the Outer Banks in Dare County, North Carolina, a remote wilderness where the airspace is exclusively reserved for military aircraft to practice targeting and ordnance releases on the targets below. Even in this small bit of airspace, the jet can do a hell of a lot, especially in the hands of a skilled pilot like the skipper. After our G warm-up to make sure that our bodies were ready to handle dynamic maneuvering for the day, we plunged into aerobatics—aileron rolls, barrel rolls, pirouettes. Before I knew it, we were calling Harry Dare—the man with a deep southern drawl who managed all the planes flying around Navy Dare—to request a panel check. Immediately, Harry's voice perked up, "Roman 11, you are cleared for a panel check with speed and altitude at pilot's discretion."

"Roger that! Roman 11 is proceeding inbound," the skipper chimed with a hint of glee as he bent the jet around and pointed our nose directly at the target area.

Harry had authorized us to fly as low and fast as the pilot deemed safe, right past the safety tower, so that Harry could look out his window and make sure that our jet wasn't missing any panels from all the maneuvering. As we leveled off right above the treetops, the skipper bumped up the throttles, and we plummeted toward the bull's-eye. I stepped the RADALT (radar altimeter—the system that keeps us from inadvertently flying into the ground) down as we kept descending and accelerating toward Harry in the tower. I won't tell you exactly how fast or low we went that day, but just know that it was hundreds of miles faster and hundreds of feet lower than I'd ever flown in a training jet. We were so low I could've read the logo off of Harry Dare's trucker hat had we not been going so damn fast.

Whoosh! We screamed by the tower and the skipper pulled the jet's nose straight up so that we looked like a space shuttle in launch. I could still hear the roar of our engines in full afterburner trailing behind the jet when Harry came back on the radio.

"Roman 11, that was a good one!" He was half yelling over the jet noise, but I could hear the grin in his voice. "Panels all intact. You're authorized to proceed as desired."

As we leveled off from our Bullseye Nose Hi maneuver, the skipper cruised straight and level, deconflicting the airspace with a section of two other jets joining us in Navy Dare.

I sat there, my body tingling, my heart pounding like I'd been injected with a massive vial of adrenaline. It was a short breather, the skipper immediately interrupting my moment of calm.

"Sorry about that, Caroline. Shall we continue?" he asked, not waiting for a response before dropping the plane into a split *S*, screaming and twisting through the sky toward the ground, bending the plane, burning a hole through the atmosphere.

Holding on to my holy-shit handles, I looked down at the fuel gauge and wondered if it could be right. *We've burned ten thousand pounds of fuel. In forty-five minutes?*

★

Christmas in Arizona was always special, but this time, even more so, because Grammy couldn't wait to introduce us to her new boyfriend, Earl.

I made my way up her driveway, carrying the gift photo book I gave her each year, stopping short to study the Buick with what seemed to be a camera strapped to its dash.

"Merry Christmas, darling." As usual Grammy had been waiting at the window and popped out of the front door before I'd even knocked. She came out on the porch to greet me. "Where is he?" she added, looking over my shoulder.

"Oh, just me. He couldn't come."

What kept the Minotaur from visiting was obvious. Christmas with my family, and especially my beloved grandmother, showed a level of intimacy he wasn't comfortable with. I understood it, but I didn't like it.

"Work," I added, and made a show of shrugging it off.

"Well," Grammy smiled, quickly hiding her disappointment. "His loss. Hey, did you see what Earl got me?" she nodded at the beautiful new car in the driveway.

"A car?"

She shook her head. "Better—a GoPro." She laughed.

Earl, I'd been told by my dad, both worshipped Grammy *and* could still drive at night. Sounded like a keeper to me. "A GoPro? Why?" I asked Grammy.

"We thought we'd take some action shots, but I can't get the darn footage onto my iPad." Grammy lifted the gift from my arms and kissed my forehead.

"Well, I'm sure I can help with that."

After the adrenaline rush of training, it soothed me to step into her warm, glowing house. I could rest and catch my breath. The only dark cloud was the Minotaur, or lack thereof. But Grammy was right—his loss. So far as I could tell, he spent the holiday alone on his couch in San Diego, watching football.

In the afternoon on Christmas day, I was curled up at the kitchen table, quietly on the phone with the Minotaur. Bing Crosby played on Grammy's old stereo and in the other room Earl held my grandmother's hand.

"You know, if you don't really care about this relationship, we can just break up," I whispered into the phone. "That way, I can find a fighter pilot or a SEAL who will actually be there for me."

There was a pause.

"You'd do that?" Minotaur asked.

"Watch me," I said convincingly, but unsure if I really meant it.

★

Naval Air Station El Centro is where the Blue Angels train each winter in between show seasons. The airfield lies in the Imperial Valley, just north of Calexico and the Mexican border city of Mexicali, in a flat, dry valley with mountains on either side. Heavy irrigation is the only thing that keeps the land from becoming desert entirely. Just beyond the gates are endless Vidalia onion and pepper fields, cucumbers, you name it. The barren landscape and arid climate make for idyllic flying conditions—blue, cloudless skies and vast airspace over bombing ranges far away from the Southern California congestion.

Surrounding the base, the local culture is infused with aviation. In El Centro, any dive bar or restaurant worth the cost of a beer has memorabilia plastered on its walls—autographed pictures of Blue Angels, layers of squadron stickers, and model jets hanging from the ceilings. Some of the watering holes even boast signed photos of Prince Harry in full flight gear, standing next to his helicopter. "He's not the bad boy the tabloids make him out to be," the local barmaids all said with girlish sighs.

As was typical for the RAG, our training in El Centro was a month long. In training we practiced our air-to-ground tactics, meaning we were dropping live bombs and practicing the skills that we would need to protect our coalition partners on the ground in Afghanistan.

Dropping bombs with close air support, or CAS, is a dangerous business. Ground forces have to trust that aviators will release ordnance on the correct targets, which is difficult, especially when troops are fighting in close proximity. It's stressful for us in the jet as well, because at three miles up, our targeting systems sometimes lack the fidelity to differentiate between an Afghani goat and a Navy SEAL pinned down in a firefight.

There is yet another danger—a bunch of inexperienced pilots clustered in a small piece of sky. It's easy to lose track of who's nearby and have a midair $160 million collision, and send four aircrew soaring in their ejection seats toward the desert floor. Thus, in addition to getting all the bombs set up and targeted and keeping your pilot safe, as a WSO, you are constantly maintaining situational awareness to make sure you don't have a collision. You're also checking to make sure that the pilot does not become target-fixated, a common problem. There are so many things going on in a dive delivery that it is easy for an aviator to become so fixated on the target that instead of shooting it, he flies into it. As a WSO, if we fly through any of the "no lower than" numbers we've memorized from the weaponeering, we'll ping the pilot with calls like "Pull up, pull up," or "Abort," adding more voice inflection the lower we go. If the jet flies into the ground, the WSO is .2 milliseconds behind the pilot, and that's not how any of us want to go out.

We were reminded of the dangers daily during our flight briefs.

"All right, guys, good morning. Looks like it's going to be another great day out there." Our class had gathered for live-bombing day and the instructor began his brief. "Let's kick it off with the safety SITREP [situation report] for the day. Who's got it?"

I stood up, my kneeboard card in hand. "Sir, I do." I cleared my throat. "Today's near miss occurred as an F/A-18 Hornet was practicing live-bombing runs during a close air support training mission. During his roll-in, the pilot became target-fixated as he slewed his errant targeting pipper in the dive. His altitude warnings were set incorrectly, and when he responded to the abort calls from the JTAC, he recovered so low that his engines hit a Saguaro cactus as his jet barely limped away."

A hush settled over the room, and I eased back into my seat.

"So a cactus was the only thing between this crew and a fiery crash." Our instructor paced. "He was lucky, but many aviators aren't. I don't want any of you guys planting yourselves in the ground like lawn darts, so today we're going to anchor down on our training rules and mitigation procedures."

It was one of the few times that no one balked about safety exercises. When you're in the air, flying near Mach 1, loaded with live bombs, there is no safety net. There's no "oops can I do that over one more time?" switch you can pull if you fly into another plane or mountainside, or if you drop a bomb that misses its target or you frag yourself. Mistakes, even in training like this, are catastrophic.

★

As our time in El Centro wound to a close, we had one last trip—Coronado Island, where the SEALs train, for the weekend. Fifteen of us students packed into two vans and headed down the hill to San Diego to blow off some steam.

The van ride down the mountain to the coast was hot, sweaty, and to be honest, smelling of farts. To make matters worse, our driver seemed to enjoy the scenic drive, going five miles below the speed limit in the right lane like a slug inching its way across the desert. It's funny, these jet pilots—these cocky, overconfident, supposed badasses who fly planes with a top speed of almost Mach 2, max it out at fifty miles per hour on the highway. White-knuckled, hands at ten and two, our driver/pilot hunched over the wheel like Grammy.

Come on, speed up, for Christ's sake, I wanted to tell him, but instead I tried to make myself small next to the sweaty dudes next to me. I was also hunkered down, texting. The Minotaur and I had broken up, but we were always in touch. There was no way I was going to be in the San Diego area and not see him, and both of us knew it.

We'd just held a ceremony called Kangaroo Court, in which we were reissued call signs. Progressing through flight training, our call signs change with some frequency, usually when you go to a new training squadron or when you do something dumb or embarrassing. A difference between the Navy and Air Force is that the Air Force generally issues cool call signs, like Viper, for example. In naval aviation we are generally given call signs that call out shortcomings or are meant to tease and generally have to do with poop, farts, or otherwise middle school–aged humor. My last call sign and the one I am most known by is Dutch. It was the name that was stenciled on my jet during deployment. And luckily for me it sounds far cooler than it was intended to. Dutch in my case stood for Dutch oven because I hated when guys would fart in the cockpit with me. It made me sick and made them laugh. Believe it or not, some pilots make farting in the jet a finely tuned sport, especially when it gets a reaction. In any case, Dutch would be given to me later in my career. During the Kangaroo Court that night in Coronado I became GotNo as in, "she's got no Johnson." The dudes found it ironic, because my last name was Johnson—hilarious.

My squadronmates were drunk and getting drunker as the bar rang with the tapping of downed shots. I was having a good time, chatting with Slushie, one of my favorite instructors, when a few seats down from us I heard someone slur, "Who the fuck is that, Channing Tatum?"

My head turned with others to see the Minotaur, striding into the bar. I glanced at Slushie, who winked approvingly. After the boys had finished their last round, we hopped on a party bus headed for San Diego's gaslamp district. Minotaur and I headed to the back of the bus to the open seats.

"Hey," a Marine from class said to his buddy as we approached. "Looks like GotNo's gonna get some tonight." He reached out to give the Minotaur a fist bump. "Semper fi, bro," he said with a drunken smile.

Long immune to these types of comments, I didn't even flinch, but I felt the air being sucked out of the bus as the Minotaur stopped.

"What did you say, dude?"

"Nothing."

"The hell you didn't. You apologize right fucking now."

The Marine sat cross-armed and silent, staring at the seat in front of him.

"No one talks to her like that. If you have something to say, come on outside with me and we'll work it out." Minotaur, outranking the guy and outweighing him by about twenty-five pounds, reached over and pulled the fellow to his feet.

"It's okay, babe," I said, restraining the Minotaur's flexed bicep. "Let's leave it." But I knew it was pointless.

"Stop being a fucking pussy, man. If you're going to talk shit, you'd better have the balls to take responsibility for it," Minotaur raged. "You're not going to get away with that bullshit toward Caroline when I'm around."

I'd never seen him so mad, the veins in his neck bulging. Despite his little legs, he towered over the diminutive pilot who, although he was also standing, seemed to shrink down.

Minotaur stepped aside and pointed to the exit. "Out," he said.

"Sir, no. I didn't mean anything by it."

"You're fuckin' right you didn't. This is your classmate, Caroline, call sign GotNo. You don't talk to GotNo like that. You don't talk to anyone like that. Am I clear?"

"Yes, sir," he said.

"Now apologize."

"Sorry about that, GotNo," the pilot mumbled, staring at the ground and slouched, wishing he were sitting on the vinyl seat.

The bus hissed into motion, and I quietly took a seat by the window. Up until that point, no one—not a single person—had stood up for me the way the Minotaur had, at least not in the Navy. Being an outlier from the majority, you learn you have to fend for yourself, day in day out. It's a burden so constant that you start to forget its weight, until someone comes along and acknowledges it, even shares it. Sure, it was awkward that Minotaur did so on the night of Kangaroo Court, in front of all of my peers and instructors, but goddamn, I was happy.

I slid into my seat, and he sat next to me in the dark. I smiled, feeling his heavy arm reach over, settle in my lap, and take my hand.

CHAPTER SEVENTEEN

★

April 6, 2012

My mom and dad were in Jacksonville, Florida, staying with my brother Craig before he left for a deployment in the Middle East. Since Craig's fridge had little more than beer and Red Bull, my mom was loading up on groceries when her cell phone rang. She saw the call was from Karen, her best friend back in Colorado.

"Hey, what's up?" she answered, but quickly her tone faded.

"Have you talked to Caroline?" Karen said, frantic.

"No, she's at work, why?"

"Call her NOW! And call me right back." Karen hung up and my mom stared at her phone.

Her stomach lurched and a wave of nausea hit her. Karen had, ironically, been married to an Air Force fighter pilot during the Vietnam War. Karen's husband had flown for the Air Force and then moved on to become a United Airlines pilot and fly for the Colorado Air National Guard. He died in a freak training accident when his jet malfunctioned and he tried to eject. Tragically, the canopy didn't release, and he was killed just like the character Goose in the movie *Top Gun*.

My mom's hand shook as she fumbled with her cell phone. It took her several attempts before she could call me. It rang, and rang. No answer. She left a message.

"This is Mom, call me as soon as you get this."

She knew I could be anywhere out of contact, in the jet flying, in a sim, or in a briefing, so she tried not to panic. She called Karen back. "I can't reach her . . . what's happening?"

"Are you near a TV? There's been an F/A-18 crash in Virginia Beach . . . and it's a two-seater."

My mom walked away from her full grocery cart and headed back to Craig's house. By that time, her phone rang off the hook, as one friend after another tried to reach her. She could barely get Craig's car started, all while trying to answer her phone, hoping one of the calls was me. She somehow made it safely back to the house and ran in, yelling to my dad to turn the TV on. There, on all the news stations, was live video of the crash. Firefighters and first responders streamed toward the giant black plume of smoke and flames that rose from the crash site, which was an apartment building three blocks from my house.

My parents sat glued to the screen as they watched the horror unfold. The squadron was confirmed as VFA-106, where I was currently training, but still no proof of life. Gradually, reports started coming out that the crew had ejected at the last possible moment while trying to steer the aircraft away from a school and had also dumped as much aviation fuel as possible. My mom called my brother Craig at his squadron. He had heard about the crash, and she told him to see if he couldn't get through to me by his channels.

Karen called again. "The station I'm watching says that the pilot and WSO are alive, but injured. Still no word if one is a female."

My parents continued to watch and flip between the news

programs, and eventually they saw one of the crew members being loaded into an ambulance. He was bloodied and injured, but it was one of the guys. All this time my mom had silently been praying over and over for the survival of the crew and the people on the ground. She was still shaken and feeling ill, and could only imagine how awful it had to have been for the families and loved ones that had to watch a similar scenario during the World Trade Center and Pentagon crashes on 9/11, waiting and wondering, feeling powerless.

Finally a reporter mentioned that a second injured aviator had been transported to the hospital—another male. Though she felt a wave of relief, she couldn't understand why she still hadn't heard from me. It would not be until a few days later that my mother would learn that when an incident like this occurs, all cell phones in the squadron are immediately put in a locked box to keep family members from hearing, or even worse, not hearing from their loved one.

Finally, after what seemed an eternity, my mother heard from me. She had tears welling up when I explained that I had been in a simulator and, like my mom, had heard something had happened. We ran out of the hangar and saw the smoke from the crash, but by then, I had no access to phones.

This proved a pivotal experience for my mom. Though she knew that our training and workups could be dangerous, she had naively assumed that my squadron's eventual deployment overseas would be the hardest time for the families, and indeed it was for most of them. Of course the danger would be heightened when we headed toward combat, but the reality was that every single time we climbed into an F/A-18, my mother had to deal with the possibility that human error or mechanical failure could take us down. For the next few years, she would never rest easy or take my well-being for granted.

CHAPTER EIGHTEEN

★

March 18, 2014; Embarked USS *George H.W. Bush*,
Departing the Mediterranean

We departed the Mediterranean and transited the Suez Canal on our way around the Arabian Peninsula, threading the needle through the Red Sea, the Gulf of Aden, and finally arriving at our destination—the North Arabian Sea—the operating area where we planned to spend the bulk of our deployment.

Throughout our travel around the Arabian Peninsula, and during our entire deployment, we constantly flew missions, trained, and practiced our tactics. There were fifty-four fixed-wing aircraft aboard the *Bush* and another twelve or so helicopters. Our air wing consisted of about twelve hundred people, 180 of whom were aviators. With aircraft taking off and landing and helicopters operating twenty-four hours a day, the flight deck, simply put, was one of the most dangerous and terrifying work environments imaginable. There are four runways on the flight deck of a carrier from which aircraft can launch. Yes, that steel cork in the ocean can launch two jets at once, with four jets launching within thirty seconds of one another. Each blasting off with thirty-four thousand pounds of thrust, which is enough force to knock over a building. The planes themselves are monstrous, heavy, and loaded with weapons.

One day, while waiting to take off for a training flight, my pilot and I had a front-row seat as one of the blue shirts—enlisted maintainers responsible for towing jets around the flight deck—almost lost his life. He went to pull the chocks from an F/A-18 right as the plane was pulled forward prematurely, and because the kid neglected protocol and improperly removed the chock, his leg was run over by the plane. As the plane rolled forward, I watched helplessly as he fell to the ground, the tire crushing his leg. His foot, in a cheap, steel-toed boot, squeezed out on the other side and popped off, leaving him writhing on deck, screaming in agony, blood seeping from his freshly amputated leg. We called tower to get the medical crews up on the flight deck, but after doing our duty to get the kid help, we had to take off. Flight-deck operations could not stop, because we had to clear the deck and make room for the helicopters to land and medevac the Sailor off to the nearest hospital in Oman.

Running the length of the carrier, there were painted lines called foul lines, and these delineated where it was safe to stand. Depending on whether aircraft were launching, landing, or hovering to come in and land, different foul lines were active, and should an aviator, maintainer, guest VIP, or admiral cross on the wrong side of the line, they were tackled out of the way by the safety observers. When doing anything on the flight deck, every person had to keep their head on a swivel, because something was always trying to kill you. There were warheads hanging off jets to duck under, and hoses of explosive jet fuel to step over. There was jet blast to dodge, and turning props that would chop you up as if with a samurai sword. All of the dangers made the space reminiscent of a medieval gauntlet. The arresting cable—which we caught with our tailhooks to help us stop on the runway—was a steel cable three inches thick. When a plane snagged it, it bent like a rubber band from the force.

On the *Bush*, there were three cables in constant use, and like all equipment on the boat, they were prone to wear. The Navy has strict maintenance requirements for how often the cables must be swapped out, but even so, cables have malfunctioned and snapped. If one breaks after a jet has slowed down, the jet cannot stop, and most of the time, it cannot fly away, so it precariously rolls off the carrier's front end. If they can, aviators in those situations eject before the jet plunges into the ocean. (E-2s do not have ejection seats, so their crews must try to fly away or ditch after the plane has crashed into the water.)

It's dangerous, not only for the crew in the plane but for everyone working on deck, as the broken cable instantly transforms into a deadly scythe, whipping back and forth in a space that's packed with hundreds of support personnel. With its speed and energy, the cable can easily cut a body in half, take off a leg, or, if you're lucky, just send you flying twenty feet, leaving you with a concussion and shattered limbs.

Even though we didn't have any cables part during our deployment, we knew the dangers, and so we spent as little time up on the flight deck as possible. When we were assigned our jet for the day, we found exactly where it was parked, or spotted, and then planned our route to it using the inside labyrinth of passageways, popping out as close as we could to the jet. This was especially necessary at night, when all of the daytime hazards were amplified by the dark and the flight deck that was dimly lit. Much like Alcatraz, if something went wrong on the flight deck, there was literally no escape. Unlike on a small ship, swan diving into the waves below is not something you want to do. Not only will your body get sucked underneath the boat and chewed up by the gigantic propellers, but the deck itself is nearly ten stories high, moving and pitching in the waves.

There are steel nets off its sides to catch anyone who might blow overboard or dive from the path of a rogue aircraft, but trust me, this is not something anyone wants to do.

On the flight deck, you cannot have any skin showing because of safety concerns, so our Sailors dressed from head to toe in long turtlenecks, heavy pants, and thick protective gear. In the Gulf, where temperatures routinely reached the 120-degree mark (and with the engine blast, often exceeded that), they worked in face masks, helmets, goggles, and gloves, drenched in oil, content to be there, working on jets and serving their country.

What most people don't know is that the average age of the men and women working in this insanely perilous environment is around twenty years old. During the Vietnam War, it was nineteen. Literally, the flight deck is manned by kids, fresh out of high school, maybe a little college under their belts; enlisted Sailors who volunteered for the Navy, having no idea what they were getting into. Some of this made us nervous, especially when an inexperienced one was taxiing our plane so close to the edge of the flight deck that the tires slipped on the greasy surface, threatening to send us plunging to our deaths. But at the end of the day, we wholly trusted these young boys and girls and were grateful for the risks they took. Our lives, in the most literal sense, were in their gloved hands.

★

March 2014; Embarked USS *George H.W. Bush*,
transiting the North Arabian Gulf

The morning of March 23 was the day I heard the command, "Launch the alert 30." Seventeen minutes later, Crocket and I tore off the end of our carrier in search of an Iranian F-4, a two-seat fighter jet. Our plane was banking hard to the heading 330, wings shrieking through the hot, humid gulf air.

Crocket had the controls. My eyes were trained on my radar looking for our contact. "Got 'em," I said as the contact appeared on my display. Almost as soon as our radar picked the plane up, the F-4 changed his course. "He's turning away." I checked his location superimposed over a chart on my other screen.

He was back in his airspace and out of our vital area, flying what I assumed to be a standard patrol route.

"Let's continue to monitor him, Crocket. Stay on our heading and wait to see."

Keeping an eye on the F-4, I bumped out the area on my radar and my heart jumped in my chest. "We have another contact," I said, seeing another bogey appear on our radar attack display.

"Let's check it out," Crocket said.

I gave him a new heading, and he arc'd the jet around, tearing into a new part of the sky. Something didn't feel right. Two encounters with Iranian fighters in one day was unheard of.

We were in hot pursuit of the second intercept, talking to controllers back on the *Bush* who were confused and not giving clear answers. The ship's controllers were thinking and operating at the twenty-knot pace that the boat was steaming at, but we were closing in on the bogey at over five hundred knots, trying to get a declaration of the unknown aircraft.

"Lion 11," I kept repeating to the controllers, "contact BRAA 305/45 23,000 declare." I was essentially asking them to tell us if they had any other intelligence to give us if the contact we were pursuing was declared anything other than an unknown aircraft operating in our airspace. But despite our inquiries, we got nothing out of them.

Closing at five hundred knots, we were seconds away from discovering what was on the horizon. And then we saw it.

"Tally one, right one o'clock high," I said to Crocket, "an Iranian P-3."

We were not supposed to join up on a P-3. The plane was just a reconnaissance aircraft. Immediately, I heard the controller in my ear telling me what I already knew, "Contact is an Iranian P-3. Do not intercept the P-3! I say again, do not intercept the P-3!"

A little late for that, I thought as the plane became visible in the crosshairs of my helmet targeting sight.

"Roger that," Crocket mumbled, peeling off the reconnaissance plane. Instead of returning to the carrier, we took up a position called a defensive counter air-cap between the P-3 and the aircraft carrier, maintaining our airborne alert, just in case the F-4 decided to come back and check us out after all. But Crocket and I knew this was unlikely, and we both shifted into a de-escalatory mindset.

Heart still hammering in my chest, two words came to mind. "Pump fake, Crocket."

"You're right about that one," he said. "Always next time."

CHAPTER NINETEEN

★

October–November 2012; NAS Oceana, Officers' Club,
Virginia Beach, VA

Late fall, as deployment neared, I was visited by a bad omen. The omen came in the form of a convenient death. After a long day of riding bikes at the beach, followed by drinks at Waterman's, I returned to my house with friends for one last round. When I went to put Wizzo, my hamster, into a play ball so we could give her a little love and entertainment, I lifted off the top of the cage and found her lying still in the paper shavings below—stiff as a rabbit's foot.

While sad, and a little grossed out, it did occur to me that Wizzo's passing could not have been timed better. She'd led a good life, as good as any pocket-sized rodent could wish, and I was leaving for a nine-month trip around the world, and let's be honest, who wants to take care of someone else's hamster? As I did not know any eight-year-olds with really, really tolerant parents, I was relieved to say goodbye to my old friend and let my classmates come to the rescue and properly dispose of both the cage and mortal remains of the faithfully departed.

Once I was officially qualified as a combat wingman, weapons systems officer in an F/A-18 Super Hornet, the Navy could, at any moment, send me around the world to join my squadron on deployment.

"In a week or two," I told Craig on the phone one afternoon, "I could be flying a fully armed fighter jet into combat."

"So you got a hunch what will happen?" Craig asked.

"Depends on the squadron."

I'd done well in all of my flight training and there were only three WSOs graduating in my RAG class, and of those three, I was the only single person of the bunch. I knew too well that in the Navy, family plays a role in where they send you and how fast they want you to get there. I told Craig, "The rumor is that the forward deployed squadron in Japan and the deployed two-seat squadron in the Middle East need WSOs."

"Middle East, as in the Gulf?" Craig said. "I'd just tell Mom about Japan."

"Yeah, well, you know how it goes. I won't know anything until after our patching. I've heard rumors the Blacklions are going to start picking people up. They have the fanciest jets, the best reputation, and some of my friends just went to their sister squadrons."

"Sounds like a perfect fit," Craig said. And it was. I dreamed they would pick me, but I wouldn't dare get my hopes up.

★

"Thanks for coming out to the class 12-1 patching, we're excited to be sending this class of newly qualified aviators out to do great things in the fleet!"

A hundred or more of us—RAG students and instructors, sixteen local fleet squadrons, parents, wives, and significant others—were piled into the O'Club for happy hour, waiting as the moderator asked the question that would determine our fates.

Like many of the ceremonies in flight school, the patching was held at the Oceana Officers' Club, but what made this event so special was that we would find out our fleet squadrons and where we would spend the next three years of our lives. We all knew the squadron—the location and people in it—would largely shape the rest of our naval careers.

The rite of passage worked like a sports' draft. One by one, each newly minted aviator was called up before those in attendance. Our class mentor first roasted each pilot and WSO separately, recalling the highs and lows of the past year, before dramatically asking the question we all waited for, and the mentor would call out the aviator's call sign.

There would be a pause for suspense before a squadron stepped up to accept the new person. Sometimes there wasn't a squadron to claim the aviator, which meant the aviator would be sent to a squadron not based in Virginia Beach. Almost halfway through the ceremony, no one yet had soaked up the west coast, Japan, or deployed billets we'd been hearing about, so we all held our breaths. An electric excitement pulsed through the O'Club. Huddled to the side of the crowd, surrounded by classmates and wives, I nervously awaited my turn. With no family in town, Minotaur in California, and my classmates chatting with family, I scanned the room, trying to distract myself when I overheard one of the wives giggle.

"It's amazing we actually look cute in these," she said, gesturing to the flight suits some of the women in the group had tailored to fit their bodies like costumes. "I mean, it's so unflattering on some of those *girls*." She tilted her head toward a small group of Jet Girls on the other side of the club. "They fit them like potato sacks."

My friend Ashley was in the group of girls wearing the "potato sacks." Ashley was a beautiful woman, a total babe. If seen in a bikini or dolled up for a night out at the clubs, most guys would describe Ashley as hot, maybe even smokin' hot. But she'd just landed from a flight less than an hour prior and had hustled straight out of her jet parked at her squadron to run over to the Officers' Club in order to make sure she was there for the start of the patching—my patching. She didn't stop to put on a dress or even a cute top and jeans, she didn't hover in front of a mirror to touch up her makeup or fix the slight hint of sweaty helmet hair, because for Ashley, attending an important moment in my life or any of her Jet Girls' lives was more important than a few strands of hair gone awry.

The exact opposite was true of the wives. All of their attention leading up to this event was focused on turning issued flight suits—our survival gear—into sexy flightsuit dresses. The tradition of spouses altering their husbands' old uniforms was supposed to be a fun part of the festivities. They went so far as to wear squadron patches with custom name tags bearing their husbands' call signs with *Mrs.* in front of them. Some kept their husband's Navy rank on their shoulder or had it changed to little cloth hearts—the rank of Navy wife—where the rank insignia should have been.

I glanced down at my own uniform, this so-called potato sack which I had no control over. Yes, it certainly was not flattering on my body. But I depended on it to stay alive. The baggy green attire was not designed to look good in a photoshoot but to keep my skin from burning off if I ejected from my jet. The lieutenant junior grade rank was not stitched on the night before, but earned over the past seven years of study, toil, and immense personal risk. My hands inadvertently clenched my drink napkin as I felt myself seething at these women who were dressed for a night of trick-or-treating as slutty pilots. I, like them, hated to see a fashion misstep, but they were out of place. It would have been one thing if they had donned their costumes and left us alone, but the fact that they openly criticized our uniforms and appearance was what really bugged me. These Navy wives had joined us as our guests, and yet a select few of them chose to mock me and my fellow Jet Girls over something so petty yet so important to us as our flight suits. I reeled in this momentary mix of anger and embarrassment when I heard, "GotNo, you're up."

I made my way to the front of the crowd, in a haze, freezing on stage as all eyes shifted to me.

"Well, we all know to lighten our load when Caroline is coming along on a det. Gotta make sure we have room for her five suitcases. I mean, what does she have in those things?" my mentor teased, and the crowd laughed. More of the same followed, but overall, I got off easy. After all, I still had a smile on my face when he said, "Now, which squadron wants Caroline 'GotNo' Johnson?"

A long silence followed. The longest of my life. For the past three years I had helped to decide my own fate by reaching number one. But my future was about to be made for me, and for the first time, I had absolutely no influence nor did I have any idea which way it would go.

The silence dragged on and I wondered if I would be chosen by any squadron. Maybe there was a mistake. Maybe I was forgotten. Then I heard, "VFA-213!"

The world-famous fighting Blacklions had chosen *me*.

An enormous cheer boomed from the crowd. Twenty aviators, led by the squadron skipper, surrounded me, replaced my old patches, and chanted "Drink, drink, drink!" A warm shot of Jager was put into my hand. I tilted my head back and drained the glass.

CHAPTER TWENTY

★

November 18, 2012; Virginia Beach, VA

A hail and bail is a party thrown to welcome the new members of the squadron and to say farewell to those transferring out. The Blacklions' hail and bail was held at our executive officer's, or XO's, house. Not far from Virginia Beach, the home was a beautiful Cape Cod wrapped by a porch with white pillars and hand-painted wooden clapboard. An American flag hung from a pole off the front of the house, snapping in the breeze.

Our XO was the second senior-most officer in my new squadron and, unbeknownst to me, a very young guy for his position. We hadn't met yet, so when I arrived to find someone who looked like one of my peers welcoming me, I gave the same warm greeting I would have to any new squadmate.

"Hey! Nice to meet you, I'm Caroline," I said, extending my hand.

"Come on in," he said, gesturing toward a house ripped straight from the pages of *Southern Living* magazine. "I'm Coma," he said, using his call sign.

I followed Coma to the kitchen where he introduced his

wife, and it suddenly dawned on me that he was not a peer but the house's owner and my new XO.

"Oh, sir," I said, suddenly switching into formal address. "Thanks so much for having us. And ma'am," I added, turning to his wife and extending my offerings. "Flowers and Ghirardelli-dipped strawberries for the party."

"I'll just leave you girls," Coma said, excusing himself.

I was unsure if I should be out mingling with the guys in my new squadron or in the kitchen prepping hors d'oeuvres with the ladies but regardless asked Coma's wife, "Can I do anything to help, ma'am?"

"You can call me Abby. And everything's pretty much taken care of, but thanks. How about you meet some of the ladies over here?" Abby led me into the next room and introduced me to a pretty blonde. "Heather, meet Caroline. Caroline, meet Heather. Please make sure she feels welcome."

"Sure!" Heather said, offering her hand and a winking smile. As soon as Abby was out of earshot Heather said, "So, we are dying to know . . . who's the lucky man?"

"Pardon?" I said, confused.

"Which one is your husband? We're always happy when a new couple joins the squadron."

"Um," I said, cheeks reddening, "I'm not really . . ." I glanced down at my hand.

"Oh, is it a boyfriend?" Heather said. "That's okay. Not everyone is married."

"Well, actually, I'm not married to anyone in the squadron. I'm *in* the squadron," I said. "I'm the new WSO who just joined."

The buttery chardonnay Heather had been sipping looked like it had instantly soured as her mouth scrunched into a circle. A weird, painful *ooohhh* sound hissed past her lips. And then, in a gesture that puzzles me to this day, she held her hand up in front of my face and turned away. She took a few steps toward the rest of the women who hadn't heard our conversation and pointed. "She's the new girl in the squadron."

"Hi, ladies." I gave my best debutante wave. "Super excited to be here," I said, cheerfully trying to lessen the awkwardness that suddenly permeated the room. I'm not sure what the women were thinking as they stared at me, but I'm guessing it was something along the lines of: *So this is the girl who's going to be traveling with my husband all around the world, for months on end, while I stay home, worrying about his death and our marriage.*

Smiling as broadly as I could and holding my shoulders back the way Grammy taught me, I turned and walked into the living room to get a beer from the trough.

My squadronmates, drinking in a tight circle, looked up timidly as I approached. I grabbed a lager and tried my best to mingle, introducing myself to a few of them, but then noticed how their eyes shifted back and forth between me and the chickenhawks peering at us from the kitchen.

I knew there would be a transition period. When a new guy or girl joins a fighter squadron, there's always a little hazing that goes on. For at least six months you're not known by Johnson or your RAG call sign, but by three letters—FNG. *Fucking new guy.*

- *FNGs should only speak when spoken to*
- *FNGs should have big ears and a little mouth*
- *FNGs should absorb every bad deal in the squadron and pay their dues before reaping the benefits*

Maybe this is part of the rules, I thought, but felt something different going on. I looked for a distraction. Like at any family-friendly potluck, there was a rogue pack of children, ranging from two to early teens. The group, like the house, was flawlessly beautiful. Exactly what you'd expect from the offspring of aviators and homecoming queens. I wandered over to the kids.

"Hi," I said to one of them.

"Hi. My name's Ben." A little boy smiled up at me. "You work with my daddy?" he asked. "You fly jets and kill bad guys?"

"Well, I fly jets just like your dad, but I haven't killed any bad guys."

"But you would, wouldn't you?" he persisted.

"Yes, if I needed to," I said, my undrunk beer warming in my hand.

"Well, that's *really* cool," he beamed and zipped off, chasing the happy swarm of kids.

For a moment, I felt purpose flood back. Ben's father was a hero and in working with his father, I was, too.

"How's the party?" I turned to see Taylor approaching, a smirk on her face. "Fun as always?" Taylor knew how these things went.

"You could probably guess, based on the fact that I'm out here with the kids." I forced a smile. I told her about Coma leaving me with his wife and about Heather holding her hand in front of my face, and how the guys ignored me after I got the cold shoulder in the kitchen.

"Yeah." Taylor slugged her beer. "Typical dudes. So afraid of their wives."

"Glad it isn't just me," I said.

"It's definitely not you," Taylor said. "Seems like it wouldn't be so hard for them to just treat us normally. Say something like, 'Hey, this is so-and-so. This is my dog. Look at my cool watch, it's fancy. Isn't the weather nice?' You know, common courtesies."

"Yeah," I agreed. "Believe me, I know."

"I'll let you in on a little secret," Taylor said and went on to explain her MO—show up late and stay for a requisite thirty minutes, just long enough for everyone to see you participated. "Then I get the hell out. I can't stand all the mando fun."

Mando fun is short for events that we refer to as "mandatory fun," meaning all of these official functions in the Navy where you have to show up and act like you're having a blast in a polite military way, of course.

When the hail and bail was finally over, I fetched my platter from the kitchen and quietly saw myself out. Taylor had already gone, so no one noticed when I left. When I got in the car, my phone buzzed in my purse.

"Hi, Mom," I said.

"Hi, honey. How does it feel to be a Blacklion?" my mom asked. She knew it had been my dream.

"Great," I said, trying to add some cheer to my voice. "Chad was there. And Taylor." My words sounded eerily like I was in middle school again, looking for companionship and affirmation.

"Wonderful. Any hazing?" she asked.

"No, nothing," I said. "I'm really lucky. This is exactly where I wanted to be. Say hi to Dad for me. Gotta go, but I'll call you later." I hung up the phone and watched the sun sink behind the marshlands.

What have you gotten yourself into, Caroline?

★

After I'd checked into the squadron, I went to the PR shop—short for parachute rigger's, the shop that maintains our survival equipment. "Ma'am, that's all we need from you right now," the Sailor said after I turned over my bag of flight gear, much heavier than when I got it in Pensacola. “Is there any gear or anything you need from us?”

It was a relatively quiet afternoon, so I replied, “Actually, yes. I hate to ask about this, but I only have two green flight suits and I’ve been wearing them for the past three years of training, so they’re pretty worn.” I shifted my weight and continued, “I’m kind of in a weird place because they’re custom-made for me and I know they take a long time to order, so I was wondering if we can please possibly order some soon? So they’ll arrive before deployment?”

“Ma’am, we don’t do custom flight suits.”

“Actually,” I said, “I have had them made for me before. It takes a while and I need them for deployment.”

The older enlisted now joined the conversation. “Ma’am, I’m sorry, we don’t do custom flight suits. We can get you normal flight suits, but nothing special.”

“Okay,” I said, deciding that these guys were going to be a pain in the ass and that I’d figure it out another way.

A week later, three junior officers who were my rank but senior to me pulled me into one of the briefing rooms. They directed me to sit in a chair in the middle of the room. They stood and began circling me.

One of the JOs spoke first. “Who are you to think you can demand custom-made flight suits when you’ve been here since breakfast? Why are you already trying to outdo our old swag and buy new stuff? Why aren’t you like the other new guys, sitting in the skiff and studying with your mouth shut?”

“I’m sorry. I’d been instructed to get new gear. I need custom flight suits and they take a while to order.”

Another JO spoke. “Need or want? What makes you think you know it all? You have a bad attitude, Johnson. You’re cocky, and you better wise up if you want to make it in this squadron.”

After twenty minutes of this treatment, I understood that they just wanted the five basic responses I'd learned as a plebe at the Academy—*Yes, sir. No, sir. Aye, aye, sir. No excuse, sir. I'll find out, sir. Sir, sir, sir.*

So I apologized and gave them what they wanted. I would try harder to be a better FNG and not cause waves.

When they released me what seemed like forty-five minutes later, I beelined out of the squadron, biting the inside of my lip so hard to keep from crying that it bled. I got in my car, slipped on my big, dark sunglasses and let the tears flow. I'd been working my ass off to fit in, to do the job I'd been assigned, to be a good WSO. I'd been meek and submissive and good at my job, but clearly that wasn't enough. *What else can I do?* I asked myself.

I wanted comfort, and I wanted help sorting through this but there was no one to call who could help. Minotaur and I had drifted apart with the distance, and I even thought he might be seeing someone new.

But even if I had someone to call, what could they do? What could they say?

I pushed my sunglasses back into place and looked in the rearview mirror, my girlish smile gone.

Just keep your eyes in the boat, I told myself. It was a phrase from Plebe Year describing the stare you see military people use when they're marching in formal ceremonies. Intensely looking straight ahead, no facial expression, no reactions to anything, totally unscathed. *Keep your eyes in the boat.*

★

Spring–Summer 2013; NAS Oceana, Virginia Beach, VA

Focus helped, but being an FNG remained difficult. Small stuff, things that normally would have rolled off my back, ate at me, forcing me to work harder.

I was put in charge of planning all of the change of command parties for our skipper and his family, so I organized a surprise going-away party for his wife and the other squadron wives that started on a decorated party bus that took them to the beach for professional photographs at sunset and a secret dinner at a local restaurant.

The evening went flawlessly, and my hard work felt well worth it until later that evening when one of the wives told me, "You know, that maxi dress is far too revealing."

Keep your eyes in the boat, I told myself.

Another night while dining with Jet Girls from another squadron, my entire squadron stumbled in on a pub crawl. Everyone—the higher-ups, all the guys—had *forgotten* to invite me.

Eyes in the boat.

Keep your eyes in the boat.

One of my Sailors had his leg amputated after a motorcycle accident. I worked with my chief to keep him in the Navy, and while he was laid up in the hospital, I bought him a flag from his favorite company, customized it, and brought it down to the maintenance shop with a couple of cases of soda for a signing party. When I returned to the Ready Room, I discovered one of the cans of soda had punctured and leaked onto my boots. I heard a couple of Blacklions in the back of the room snickering and was harassed for weeks about the incident.

Eyes in the boat.

When I tried a new salon and my hair came out a little lighter than usual, my boss joked in front of a group of peers, "Whoa, looks like you're going to star in a film later tonight," and then referenced a well-known porn star.

Fucking keep your eyes in the boat.

While working at my desk, a JO spilled his drink and it leaked onto the papers I was working on. "Hey, can you please grab a paper towel and clean that up and not ruin my hard copies?" I asked him.

"Fuck you, new guy," he said. "Cleaning is a woman's job."

Fuck you right back. *Eyes in the boat.*

Nothing in my background had prepared me for this bullying, much of it subtle and occurring behind closed doors, leaving our senior leadership either unaware or able to ignore what was going on. None of it was outwardly criminal, but the microaggressions were constant and relentless. Recounting these instances makes me uncomfortable, and I wish I did not have to include them in this book. But I also wish they didn't happen. If they hadn't happened, I'd still be in the Navy.

These incidents were just a few of a hundred. Many were stupid, slight, and normally ignorable, but they occurred in a stressful environment where I already felt like the odd man—woman—out. Some so minute and meaningless they shouldn't have even registered as real issues, but building up every day, multiple times a day, over time the pile grew into a wall I could not climb and a weight no one should have to carry.

★

Summer 2013; Virginia Beach, VA

As my fellow aviators began to marry and have kids, I watched familial stresses popping up like mushrooms after a good hard rain. Some guys found a way to manage it all, breaking away from the squadron when they could to be there for their wife and kids, and others doubled down on work. Rather than going home to a pissed-off, pregnant wife, they stayed later and arrived earlier, pissing the wife off further.

When families and the squadron got together for mando (and non-mando) fun, the more time their men invested in work, the harder the wives would stare, studying me with suspicion and apprehension. I get it. After all, I was spending fourteen hours a day with their husbands, in close contact, day in and day out all over the world with weekend ski trips and drunken rambles.

Dealing with icy stares and icier introductions from the wives was irritating, but par for the course. What I didn't expect, what I found disappointing, and even distressing, was how the guys—who spent enough time with me to know my motives—came to treat me in front of their spouses during the now much-loathed mando fun.

Case in point was the night our squadron held a *Miami Vice*–themed potluck. Never missing a chance to don a costume, I went all out. I wore a Betsey Johnson dress bursting with neon florals, accented with huge earrings and rockin' heels. I even bought a hair crimper, and with waffle-ironed locks pulled up in a bow, I looked like I'd time-traveled right out of the eighties.

The party was at the house of a department head, another beautiful mansion right on the water. Arriving right on time, I parked my car on the street and looked in my rearview to see two other couples pull in behind me. I got out of the car, smiling and waving.

"Guys, hi! Could one of you maybe give me a hand?" I popped my trunk and tried to pull out the sodas and bulky ice bags I'd been asked to bring. "Hope I got enough," I said, talking into my trunk. I looked up, crinkled hair falling in my eyes, to see that they'd already strolled past.

I'm sure they didn't hear me, I told myself, hefting the party supplies and falling in step three strides behind them. They climbed the front stairs, and the aviators held the door open just long enough for their wives and then quickly moved inside.

"Hey, can somebody hold the—" I hustled up the stairs, tottering on my three-inch heels. I tried to catch the screen door with my neon-clad foot, but it snapped in my face.

I took a step back, set the ice on the ground, and swung

the door open. I walked into the party, bag marks on my arms and water dripping down my dress, my smile faded. The guys who had just entered before me were the most senior of the twenty junior officers in our squadron, and as I stepped into the kitchen, I noticed a cluster of more junior aviators—guys who were brand new to the squadron—looking my way. It was clear from the pitying expressions on their faces they felt sorry for me. Not wanting pity, only respect, I plastered on a smile, unpacked the spread at the wet bar, and ducked into the bathroom.

In the Navy, we model our behavior after our instructors and the senior leaders who set the example for their subordinates to emulate. In this instance, the senior guys weren't being outright sexist, making lewd comments or pointedly making fun of me as they had in the past. But in some ways, it was worse. They sent a clear message down the chain of command: Caroline is not someone to wait for, not someone to help. She doesn't deserve the common courtesy you would give a stranger following you into a gas station.

When senior leaders perpetuate that mindset, it takes someone truly brave to break it. Someone must step forward and say, "Guys, that wasn't cool. She's one of us."

In the Navy, and in the military in general, it is a rare person who will become an ally and stand up for someone else and risk becoming the pariah themselves.

I stood in front of the bathroom mirror, dabbing the corners of my eyes so my winged liner wouldn't run. I drew a deep, calming breath, and again rolled my shoulders back. *Remember,* I told myself. *Keep your eyes in the boat.*

CHAPTER TWENTY-ONE

★

July 2014; Embarked USS *George H.W. Bush*, North Arabian Gulf

Many of the challenges I faced in the squadron were addressed in books that Admiral Bullet recommended to me, and my mother would buy and send. I wanted to tell him about the constant scrutiny I felt at times, the seeming obsession with revealing my flaws and a growing sense that within our squadron and in naval aviation as a whole, there existed a culture that was motivated by exploiting flaws in others as a way to bring them down. The reason why we so rigorously critique is to become better, but something was poisoning this ethos. All around me I saw people who used the minor shortcomings they found in others to their advantage. When you're building a team, fear of failure can be incredibly motivating, but when you see others trying to capitalize on your failures to get ahead it is another thing entirely. You no longer seek to be better but to hide mistakes. This type of thinking is absolutely deadly in a jet.

Our talks often veered into leadership and though he prodded me about the obstacles I faced, never wanting to abuse our relationship, I held my tongue. I knew the admiral passionately loved the Navy, and was always asking what else it could do for me and for women in its ranks. I never fully opened up about the problems I'd witnessed festering in our squadron, and to a degree, in the Navy in general. Even after all the years, Bullet had never become disenchanted with the system, and so I found it hard to be critical of his Navy.

But there were instances where I saw improvement. The Navy is capable of course-correcting senior leadership when it involves junior officers. I've seen it firsthand. Take, for example, the case of a WSO we'll call Cupcake. Cupcake was a straight-up rat bastard to me. He joined our squadron midway through deployment, rotating in for a senior WSO who'd rotated out. He hadn't had the best reputation in the RAG as a student, but we needed a body, so they sent him in.

When he arrived, his personality and character were immediately transparent. When someone in the aviation community joins the boat from ashore, they often bring food. Or I should say, they always bring food. Junk food and treats we don't get on the boat, and like locusts, we descend and eat it all. After you've been at sea for months on end, anything from the mainland, from any culture, so long as you can eat it, is greatly appreciated. Everyone knows this.

Cupcake showed his colors the very first day, stepping off the carrier onboard delivery plane and presenting his gift for the Blacklions—a single sleeve of off-brand chewing tobacco. It wasn't even something like Skoal or Kodiak or Copenhagen, but the kind of cut-rate snuff that smelled so disgusting that a single whiff of it would make even a casual dipper gag. The proud smirk on Cupcake's face said almost everything we needed to know. He'd brought the gift for himself and those like him, but he cheaped out in such a way that only those with bad teeth and sour rot guts would really be able to enjoy it. He had what I liked to call sling-blade swagger, and he strutted onto the boat with the air of an uppity moron.

Immediately, I had problems with Cupcake. During his first day of duty, I had to communicate with him on the phone from the tower. I was standing next to Airboss, who, as his name implies, is boss of the planes in the sky. Cupcake and I were taking orders from Airboss, supposedly working together as a team to troubleshoot a problem with a squadron jet, but he apparently decided he didn't like the information I was relaying, so he hung up.

Stunned, I called back, thinking we got disconnected, and he hung up again. I called our squadron's maintenance master chief to circumvent Cupcake and resolve the problem. The crisis was averted with help from the master chief, but for the rest of my watch rotation, I kept wondering what the hell happened.

Maybe he doesn't like that I'm a woman, or that I'm senior to him? Maybe he doesn't understand that a direct order from Airboss isn't elective?

I'd never experienced this kind of insolence. I didn't even think as an officer you could hang up on someone, especially someone senior to you who you're supposed to be trying to impress. I decided that rather than let the problem simmer, I would address Cupcake directly, so I found him in the Ready Room after the crowd had thinned. "Hey, man, can we chat? I'd like to just understand what happened. If I did something wrong or if there was a miscommunication, I don't want it to happen again."

He turned to me, bottom lip so packed with brown goo it looked like a goose turd.

"I'm not fuckin' talkin' to you," he said, dropping himself down in a chair like a child refusing to leave the playground before he was ready.

"I'm sorry?" I tried to stay calm.

"It's your fault for fuckin' talking down to me," he raged. "Don't think you can boss me around, you dumb cunt."

If this had been a cartoon, there would have been steam coming from my ears. I wanted to throttle the bastard, but I knew I had to walk away before I absolutely lost control. I marched to the senior-most JO's room and banged on the door.

"What's wrong?" Sully asked, seeing my bright red face. "Sit down, Caroline, sit and tell us what's going on."

But I was so mad, I had to pace.

After I told them everything, they were furious. "This guy, he's what? Been here for two days?" one of the other JOs said. "This is ridiculous."

"Agreed," said Sully, who looked at me with clear blue calming eyes, "but don't worry, Caroline. Leave it to us. We'll handle him."

They tried to talk with Cupcake, but no surprise, they also failed to reach him. Even Sully, who all the guys loved and looked up to, could not get the message through to Cupcake. The guy had such an offensive personality that he rapidly became the odd man out. No one wanted to sit with him, eat with him, fly with him, and he slowly became ostracized from the people he should've been closest with.

I'll admit, part of me was satisfied. The guy was a jerk and jerks get what they deserve. But still, while I was experiencing a little relief from the constant abuses (perhaps due to the distraction of deployment), I'd experienced cruelty firsthand, so I couldn't help but feel a twinge of sympathy when I saw Cupcake by himself, batting his dip tin across the empty table like a hockey puck.

It seemed nothing was going to break the cycle until we got our new XO, Chick, who was Coma's replacement. Being the leader that he was, Chick noticed Cupcake's isolation right away, assigned him duty in another part of the ship, and called a meeting with the rest of the JOs. He wasn't going to let someone get left behind.

"Listen," Chick said. "We know Cupcake is acting like an asshole and he's not listening. If you want him to be better, you're gonna have to do the opposite of your instincts. Go out of your way to pull him and make him a part of the team. I know you guys don't want that, and maybe he acts like he doesn't, either. But that's what's right for him, that's what's right for our squadron, and that's what's right for our country. It will serve us well in battle."

And so, every single one of us changed the way we treated Cupcake. We invited him to sit with us when we ate, we talked to him, we joked with him, even complimented him when he improved in the jet. While Cupcake's performance didn't do a complete one-eighty overnight—it improved marginally—his attitude completely reversed course. Cupcake went from being miserable and alone to apparently content and a welcomed part of the team. It seemed that he couldn't have been happier, and I was happy for him.

CHAPTER TWENTY-TWO

★

October 2013; Naval Air Station Fallon, Fallon, NV

Picture Burning Man's flat, inhospitable, barren desert landscape. An hour away is Fallon, Nevada, with its downtown of strip malls, dog-eared casinos, empty storefronts, liquor stores, and diners. And just outside of town you will find Naval Air Station Fallon, home to the Naval Air Warfare Development Center, or NAWDC, where the most elite aviators and instructors develop the tactics that keep the United States ahead of our most advanced adversaries and where the art and science of dogfighting is perfected. NAWDC is also home to the United States Navy Strike Fighter Tactics Instructor Program, or SFTI, popularly known as Topgun. What the instructors and students at Topgun can do with aircraft is simply jaw-dropping. And it's no wonder Topgun's home is Fallon, because there is absolutely nothing to do out there but fly.

Deployment is when military units depart their home base and establish themselves overseas to engage in combat operations or international exercises and patrols. Naval aviation squadrons are constantly either on deployment, winding down from deployment, or preparing to go on deployment, a phase of training called workups. During workups, the Blacklions trained in different locations, constantly on the road, bouncing from Key West to home, to Fallon, to home, to Fallon, to home, to the boat, to home, to Fallon, to home, to the boat, to home, and, finally, deployment.

Not only were the ranges out west best for sharpening our skills and testing our strike tactics—dropping bombs, close air support, air to air combat—they provided the perfect training ground for Afghanistan—tall mountains, high elevation, lots of sand. Even the harsh weather was the same. At Fallon, our flying elevated to a whole new level. We powered through long days of massive war games, followed by four-hour debriefs in front of our entire squadron plus the six other squadrons in the air wing, and the highly trained SFTI instructors. The stressful and competitive work, combined with long hours, made blowing off steam on weekend trips or in the O'Club essential.

One day at our workup exercise called Air Wing Fallon, I'd flown a long and complex, multiplane air battle and endured a four-hour debrief. Exhausted, hungry, and ready for a cold beer, I realized most of the squadron had already left for dinner. I showered, changed into a pair of jeans, a comfy Lululemon shirt, and a down vest. With the sun setting and the brisk fall air setting in, I hurried across the officers' quarters courtyard and ducked into the O'Club.

The Fallon O'Club, probably my favorite bar in the world, is filled with memorabilia—gifts from each Topgun class, ranging from propellers to a jukebox, and the aviation history is so rich, you can almost hear Goose and Maverick singing "You've Lost That Loving Feeling."

Since it was still early in the night, a few seats sat open at the bar, so I took an empty stool next to a guy by himself. Like me, he wore jeans and a warm sweater, and seemed to be minding his own business. Just the kind of companion I was looking for after a long day.

"Get you something to drink?" the bartender asked. "On tap we got Miller Lite, Budweiser, PBR . . ." she rattled off the beers.

"Don't have Stella, do you?"

"Actually I do. You and that fella are the only people who've asked for my Stella all week." She poured my beer, tilting it just right so as to avoid a foamy head.

"Good choice," said my neighbor in a slightly highborn British accent. "Can't stand that watery American shit." He had the kind of voice you wanted to bottle up and drink, but I wasn't gonna let a Brit trash America, even if he was right about our flavorless beer.

"English beer is shit, too. Apparently you agree, since you're drinking a Belgian."

He perked up, arching an eyebrow. "Now, now, don't jump to any conclusions." He smiled. "You think I'm English, don't you?"

"I know you are," I said, sipping my perfectly poured Stella Artois.

"Only half English. My mother is German."

Calling his bluff, without pause I spoke to him in German, "*Wirklich? Wenn sie echt Deutsch ist, wo kommt sie denn her?*" Really? If your mother is truly German, tell me, where is she from?

"*Aus Bayern, naturlich,*" he replied in perfect German. She's from Bavaria, of course.

At that moment, if he had stood up and offered me a hand, I would have walked with him down to the flight line, stolen a plane, and flown our F/A-18 to a twenty-four-hour wedding chapel in Reno. In a few short sentences he had me, and he knew it, too.

We sat and talked into the night. And the best part was we didn't talk about flying. No shoptalk of tactics or training, we didn't even mention the military or airplanes. Instead, he described his favorite villages in Germany. We talked about mountains in Austria, where to stay in Dubai, even literature. He actually read books, and fiction, too! Real novels without airplanes in them. For an hour and a half, the rest of Fallon disappeared, and it was just the two of us, chatting, laughing.

After dinner, he waved for the check, and looked at me slightly shaking his head. "Who are you? Who's this girl sitting alone in the O'Club speaking German and having dinner with a guy she's just met?"

And I threw it right back at him. "Who are you?"

"You don't know me?" he said, smiling, arrogant. "Surely you know who I am."

"You may think you're special, but news flash—not everyone knows or cares to know who you are," I said coyly. "If you're so important, though, please enlighten me."

As it turns out, the unassuming guy with the good beer taste—let's call him Burberry—was the first British pilot to ever attend Topgun, which would be the equivalent of the first British person to join the NFL and go on to win the Super Bowl in his first season.

Just then, the rest of the squadron, having finished dinner, showed up half drunk and looking to get all the way there. The room morphed from quiet to raucous, and Burberry excused himself to go study. "Hope I see you around," he said.

I drifted over to my squadron, chatting with the guys who already knew Burberry by reputation and were in awe. Had it been any other guy with me at the bar, I would have been pelted with questions—"What were you doing with that guy? What was he doing in our O'Club?" But since my companion was a famous British pilot, one of the eighteen current Topgun students, the chatter turned to the elite flight program and how to get in.

The guys might have missed it, but Taylor sauntered over, grinning knowingly. "Oh my God!" She pulled me aside, smelling of vodka and citrus. "You and that Brit. I could feel the sexual tension from across the room. You've got a thing going, don't you?"

"Not yet."

★

Burberry and I would have an intense whirlwind relationship. The kind that happens when both parties aren't concerned about the future, but let go and discover something previously unmatched in other partners.

As with most military relationships, timing was perfectly terrible for anything long term. Burberry would be leaving Topgun for Virginia while I would be on my nine-month deployment to the Middle East. We both knew what we had would burn brightly, but only for a short, intense time. But for once in my life, I lived fully in the moment. All fun and no stakes, so long as we kept the relationship away from the squadron. Even though Burberry was totally within bounds for dating—he was not in my unit or chain of command—it still wasn't acceptable for a woman to have any kind of romantic escapade, unless it ended in a serious relationship or marriage. While guys were lauded and praised for relationships and trysts, women in the fighter community were shamed. Called slutty or mocked for exploits. So in keeping with the survival protocol I'd developed for myself, I kept my cards close to my chest, and kept the relationship a secret known only by my closest friends . . . until now.

CHAPTER TWENTY-THREE

★

March 11, 2013; NAS Oceana, Virginia Beach, VA

I sat in the Blacklion Ready Room, drinking coffee and preparing to fly, when one of the guys got up from the table and stepped away to answer his phone. He came back, pale, and struggling to speak.

"A prowler," he said, "just crashed . . . in . . . uh . . . Washington. Two hundred and fifty miles from Whidbey Island. Unknown at this time if there are any survivors."

Someone grabbed the remote and switched the TV from *SportsCenter* to FOX News. We read the scroll at the bottom—*Navy Jet Crashes in Central Washington*.

The prowler, an antiquated and notoriously dangerous plane to fly, is currently being replaced by the E/A-18G Growler, an electronic attack version of the F/A-18, meaning instead of dropping bombs they jam the enemy's radar and comms. Prowlers are usually manned by three or four aviators: one pilot and multiple naval flight officers. Naval Air Station Whidbey Island is just northwest of Seattle and I thought about who I knew there.

My mind immediately flashed to Valerie Delaney, a prowler

pilot I knew from the Naval Academy. She had been the first female midshipman to study at the French military academy, Saint-Cyr, while at the same time I had been the first to study at the German naval academy.

In our community, when a tragedy like this happens, we refrain from talking about who it might have been until we know for certain, but that morning, I had a strong premonition it was Val. As was the case with the training crash the year before, every officer in the mishap squadron would have to lock their phone in a metal ammo can, ensuring no one could call their families until the loved ones of the deceased had been notified. When someone perishes in an accident, depending on their wishes, normally the chaplain and casualty assistance calls officer will show up at the victim's house to notify the family in person of the loss.

This happens rapidly. As I had thought, it was Val. Once Val's family was notified, the word had spread by six p.m. that evening. Val had been at the controls, flying the VR-1351 in a low-level training route known as a million-dollar ride, the most challenging run for prowler pilots based in Whidbey Island.

I was living with my friend Ashley, who, like me, had become an F/A-18 WSO. We lived in a gorgeous home she built for herself in Virginia Beach. Ashley came home from the base close to midnight that night. I lay in bed sleeping with her puppy but woke when I heard the door unlocking.

"Dad, it was Val. Valerie Delaney, one of the Jet Girls," Ashley sobbed into her phone to her father, a retired Navy submariner. "I just can't believe it. It was one of *us*. Could have been me."

I let her puppy slip out from the covers and followed downstairs to comfort my friend, but before I made it to the staircase, I heard the door shut and lock, followed by the rev of Ashley's BMW starting. She left as quickly as she had come in.

Normally, I might think her distress was a bit of an overreaction; after all, she didn't even know Valerie that well—neither of us did—but it was more that Val was one of us. She was a Jet Girl, and suddenly, she was gone

Hey, Ash, I heard you come home. I guess you just got the news, I texted.

On my way to Taylor's, Ashley wrote back. *She's a wreck and I need to be w/ her for the night.*

Taylor and Ashley were best friends, and from the sound of it, Taylor, normally an emotional rock, had been badly shaken. I tried to compartmentalize the crash so I could sleep, telling myself, *It was an isolated incident, we can't know what happened, stay focused on tomorrow.* I eventually coaxed myself into a half sleep, but for the rest of the night, each time my brain sunk into deep sleep I awoke in a nightmare, seated next to Val. Ahead of us was the ground growing larger and as it rushed toward the cockpit, I saw the slant of the wide field we were flying into. I felt the heavy pull of the Gs in the turn. I tried to yell to Val to help her, to alert her to the threat but couldn't speak. The cockpit crumbled, flames enveloped us, and I woke up in bed, believing it was on fire. Each terror jarred me awake, my heart hammering and my pillow soaked with sweat.

In a turn, flying as fast and low as Val had been, you only have six seconds before you impact the ground. Six seconds. It really could have been any one of us.

In fact, it was three of us.

Three people died in the crash, all immediately upon impact.

After waking up repeatedly from the nightmares, I took a shower, made coffee, and drove to the squadron in the predawn darkness to brief for my flight. It doesn't matter what your emotional landscape is, the show goes on.

Seated in the Ready Room with my usual cup of coffee steaming beside me, I mentally worked myself into a hard focus.

Visualizing my flight, I walked through my checklists, seeing my fingers pressing the proper switches and buttons, reciting my targeting calls. This process of preparation is called chair flying, and it is something aviators do to perfect their craft. Blue Angels go through it before every show. And that morning I needed the time and focus to stay safe.

While in this state, a friendly enlisted Sailor walked into the Ready Room to empty the trash and mop the floor. He looked at me and other aviators sitting around drinking coffee, unaware that we were prepping for a simulated battle to take place over the Atlantic at five hundred miles per hour. The Sailor was jittery, like he'd had too many energy drinks. "Sir, did you hear about that crash out in Whidbey?" he asked one of the officers seated by the coffeepot. "Sounded gruesome. Three dead. A girl, too."

The pilots and WSOs who knew Val and the other victims eyed each other across the table as the Sailor continued. "I heard it was pilot error. Do you know what happened?"

I wanted to grab the kid, tell him to have some decency and common sense enough to be quiet and let us work. We all did. But he was a young Sailor working as support staff, and like a civilian, his misunderstanding of our work and culture couldn't be held against him.

"Yeah, buddy," the officer said, filling his mug, his tone clearly trying to shut the conversation down. "We know. Why don't you come by after we clear out. Okay?"

The Sailor left, maybe he got it. Maybe not. I sat back in my chair, refocused, and mentally returned to my plane.

★

Ashley's picturesque home sat on the north shore of Virginia Beach, in a neighborhood of beautiful houses rumored to have been styled after the movie *Pleasantville*. They were new, but built to look like old, classic beach cottages with broad porches, all clad in pastel colors. On warm days, neighbors would sip cocktails and play croquet on a neatly manicured croquet pitch a few blocks away.

I was out for an afternoon jog in this serene environment when my phone buzzed in my windbreaker pocket. I checked the caller. It was Eric, an old friend who had attended the Air Force Academy and lived with my family on the weekends with another cadet, Tyler. I slowed to take the call.

"Hi, Eric! How are you?" I said, winded but chipper.

"Eh, not so good," Eric mumbled and then paused.

I stopped moving. "What happened?" I said.

"I hate to be the one to break the news," he said. "But Tyler . . ." he barely spoke his best friend's name before his voice cracked.

A year ahead of me, Tyler Voss had already deployed to Kyrgyzstan. He flew the KC-135, one of the giant refueling jets that extended the range and presence of our air assets in faraway, hostile places. His plane was the kind that I would hook my fighter jet up to, to refuel in combat.

On the morning of May 3, 2013, Tyler took off from Kyrgyzstan bound for Afghanistan when a catastrophic mechanical failure caused the tail of his airplane to rip off. There was no saving the massive jet, fully loaded with aviation fuel. Tyler and two other crew members perished in the fiery wreckage.

As with Val, I'd actually heard the news of the downed plane and even suspected I knew someone on it, but then tried not to think the worst. It's eerie, once you get into the aviation community, every aircraft that goes down doesn't carry faceless service members, but friends, acquaintances, and classmates.

I hung up the phone and called my brother, then my mom. Too distraught to finish my run, I walked home, dragging myself through the front door into the kitchen where Ashley stood at the counter arranging vegetables on a platter for a dinner that night. A beautiful spring evening, the sun hung low outside the kitchen window and a breeze blew in from the bay, fluttering the flag on our neighbor's porch. I grabbed a glass of water and a bottle of rosé from the fridge.

"How was the run? You look pretty flushed," Ashley noted, lining up carrots. Friends would be arriving soon and I had to tell her before they got there. Ashley, too, had known Tyler.

I downed my water, poured two cold glasses of wine, and slid one over to her. "Hey, Ash," I said trying to keep it together. "Remember Tyler Voss?"

She smiled, trying to bring back a memory. "Yeah, of course. Cute Air Force boy from your parents' house when we were still at Academy."

Her eyes lit up, likely remembering the night she and Tyler had met—flirting by the bonfire at my parents' house for the Air Force–Navy game six years earlier. "What's going on with Tyler?" She looked over at me and I could tell from her face that she'd connected the dots.

"Don't tell me . . ." She set down her wine and leaned on the marble counter. "He was a pilot on the KC-135."

★

Sitting in the middle of the chapel at Arlington National Cemetery, I looked behind me and saw that the back of the large chapel was standing-room only. Among the hundreds of mourners, fifty-five female aviators.

A horse-drawn carriage carried Val's remains to her gravesite. Following three volleys from the honorary Navy firing squad, four jets buzzed overhead in a missing man flyover.

The military fanfare was honorable and fitting, but one of the most lasting tributes to Val came when we gave her our wings. Since there wasn't a coffin, long strips of green webbing—the same technical, military-grade fabric that holds our flight gear together—were decorated with the gold flight wings of over two hundred female aviators. These wings, technically called US Naval Aviator badges, were bestowed on us when we graduated flight school. Each measured about an inch from top to bottom and three inches wide, and stacked on top of one another, on the four-inch-wide strip of webbing, the wings for Val stretched over twenty feet. Many of the women who sent their badges didn't actually know Val, but had heard of her stellar reputation and her tragic sacrifice. I watched through tears as the gleaming wings were presented to Val's family at Arlington that morning in June.

The wings are now on permanent display as part of the Women in Military Service for America Memorial. Inspired by Val's impact, her family launched the Wings for Val Foundation, a nonprofit dedicated to promoting and supporting women in all fields, but especially aviation, and inspiring future generations of female leaders. To learn more about Valerie Delaney and her legacy go to www.wingsforval.org.

Our business is risky. It doesn't matter if you're employing weapons in combat, getting shot at by the enemy, or training to fly the war machines, it can all end in mistakes and accidents that take fractions of seconds to occur. A lot of people will glamorize death and sacrifice, but that's not my intention. I dearly wish Tyler and Val and their fellow aviators were still here with us, but you cannot defend a country without being willing to risk your life so that others may preserve theirs. To be willing to do this, to be willing to place yourself in harm's way for the safety of others, I believe is life's highest honor.

It was impossible, after Val and Tyler's deaths, not to reflect on my own mortality. In the ceremony, the priest talked about Val's motto, "adapt and overcome," which I couldn't help but apply to my own situation, as I struggled with harassment and bullying from those I would go into battle with. We were all willing to pay the ultimate sacrifice for strangers, yet some couldn't treat a fellow squadronmate with dignity and respect. I did not yet know how I could continue to endure the disrespectful treatment, but saying goodbye to Val, I knew one thing: I would adapt and overcome.

CHAPTER TWENTY-FOUR

★

May 2013; NAS Oceana, Officers' Club, Virginia Beach, VA

Just like in flight school, at some point during workups I settled into my groove and began excelling. In a squadron, aviators must advance through different training syllabi to qualify as a flight leader and beyond, so I constantly reached for the ladder's next rung. Because of the aggressive schedule to ready us for deployment, I finished my first set of quals quickly and worked toward becoming a combat-section lead. This meant I would be in charge of two jets and four people, and once qualified, I'd lead a section of junior aviators into combat with live missiles and bombs. The qualification was a big deal, requiring lots of extra study, practice, and preparation for my flights. I discovered I had a knack for briefing and breaking down the complex concepts in a way that junior squadron members could understand. Truthfully, it was kind of fun, teaching the new guys like I'd wished the senior Blacklions had done for me. I was getting positive feedback for my performance and was feeling more confident in my WSO skills.

In my ground job, after successfully completing my first two collateral duties, my superiors informed me that I'd done better than my predecessors. It felt good, after so many critiques, to gain a bit of affirmation, even acceptance. In addition to my ground job, I was tasked to co-coordinate the Blacklion Bash, a fundraiser our squadron hosted to support the Wounded Warrior Project and Special Operations Warrior Foundation. As I'd done in the past, I threw myself into event planning, ensuring everything was perfect right down to the cocktail napkins. As a result, the Blacklions raised more than $10,000 to donate that night. Toward the end of the bash just as the silent auction was about to wrap, Dianne, a helicopter-pilot friend from flight school, arrived to drop off a donation from her squadron.

"Looks like everything went perfectly." Dianne raised her beer and motioned to the packed house egging the band on for one last song. "You know he would be proud of you."

She didn't say his name, but we both knew who she was talking about. Dianne had been Minotaur's flight partner in flight school. It had almost been a year since we'd been together in Coronado. He had a new girlfriend. With all the stress of loss and the transition to Blacklions, I didn't feel like I could carry another emotional grenade, so for the first time, I was really okay with moving on.

Still, I didn't want to talk about him, so I ignored the reference. "Yeah, thanks," I said.

"So, any new men in your life?" Dianne ran her hand through her hair, a beautiful engagement ring and shiny wedding band catching the light.

I shrugged. "Eh, I haven't had time. I've been slaving at the squadron and haven't gotten out much," I lied, keeping Burberry to myself.

"Well, our friend is getting through it. Thank God."

"What?" I said. "Back up. Getting through it? What do you mean?"

"Oh—" Dianne put her sparkly hand to her chest. "Wait, you didn't hear . . ." She trailed off.

My cell phone dropped to the carpeted floor with a thud.

"What, Dianne? Did something happen to the Minotaur?" My mind flashed to Val and Tyler, and how bad things happen in threes.

"Oh gosh, sorry. He's pretty beat up, but he'll be okay. He was in Afghanistan—" She stopped to take a sip of her drink.

Dianne, like many helicopter pilots I knew, talked like she flew, pulling into a hover and pausing midsentence.

"You're killing me! Where the hell is he? Germany, in surgery? Back at home? What happened?" I tried not to grab the collar of her dress. "Start from the beginning, and no stopping."

"Sorry. I'm just surprised you don't know." She shifted her beer to her other hand. "Well, about a week ago, Minotaur was flying a mission in Helmand Province. As they were coming in to land, the copilot crashed the helicopter. His whole crew made it out of the wreck, but just outside their base, they were ambushed by the Taliban. He kept his cool and saved the day, but he's in rough shape now."

The silent auction coordinator came over. "Caroline, we need . . ."

"Not now," I said giving her a sharp, this-is-important glance.

"We've emailed some," Dianne continued. "He's dealing with it, but still struggling. You should contact him. Bet it would really cheer him up."

I arrived home well after midnight, slipped off my heels at the door, and went straight to the kitchen table to flip open my laptop.

"How did it go?" Ashley called from upstairs.

"Oh great," I said. "Tell you all about it in the morning." My email opened with a ding and I sat there, watching the blinking cursor, wondering how to begin.

★

Summer 2013; Virginia Beach, VA

A steady stream of data packets began following back and forth halfway around the world. Minotaur told me that physically, his back bore the brunt of the crash, and ironically (considering my nickname for him) he was almost an inch shorter from the impact. While thankful to be alive and able to walk, his ego and self-confidence had been shattered.

It could have been prevented, Caroline. I keep thinking that I should have stopped him from crashing.

He was talking about his senior copilot. That was Minotaur's way—taking too much responsibility in the air and not enough everywhere else.

The safety review board would convene and clear him of any responsibility so he could resume his flying, but that would take a while. I tried not to bother him with my own struggles with the Blacklions, keeping our conversations light—family and travel and joking around. He'd cheated death, and I could tell that had jarred something awake in him, but on a practical basis, he was stuck in Afghanistan for another few months, and since we had already failed in long-distance dating with three thousand miles between us, I wasn't about to try it again at seven thousand. I told him, "I'm not ready to try again. And I don't think you are, either."

CHAPTER TWENTY-FIVE

★

Spring 2014; Embarked USS *George H.W. Bush*, Gulf of Oman

We arrived in Afghanistan in late March 2014. Initially, we were there to support Operation Enduring Freedom, the war on terrorism which the United States had been fighting against the Taliban since 2001. When I say we arrived, the only people who actually went into Afghanistan were the aviators. The boat was parked south of Pakistan in the North Indian Ocean, and each day it took almost two hours to fly from the aircraft carrier to our assigned area of operations where we would get to work and then fly home.

We fell into a steady pattern of long days and short weeks. As a WSO, I flew more often than my pilot peers because there weren't any restrictions on the events I could fly. That meant I flew every day except for Sundays, which were our no-fly days even though helo pilots still had to fly. On Sundays, the flight-deck crews would rest, and the supply ships would pull alongside us in the morning, and with the help of our helicopters, the carrier would restock with fresh food, new mail, and more jet fuel for the week to come.

During my six days of flying per week, two or three of the

days I would fly into combat and the other days I would fly training missions or tanking flights, which were quick and easy. We didn't fly "feet-dry" or "over the beach" (both terms for flying in combat) every day because it was so mentally and physically taxing that our bodies needed a day off in between to be able to handle the stress. Also, we were limited by aeromedical regulations in how many hours we could fly in a month. Even with our regular flight schedule, we still had to get waivers from the flight doctor to allow us to stay in the cockpit. Like Cinderella, when the clock struck midnight—or in our case, sixty hours—time was up.

Each day I flew in-country, depending on if I was on the early, mid-, or late cycle, I would get up at least an hour before the brief; brief three hours before the flight; eat a heavy meal to hold me over; pack my helmet bag with charts, water, snacks, and magazines; get dressed in my flight gear; load my gun; grab my classifieds; and head up to the flight deck. We then had an hour to preflight the jet—start the engines and check all of the systems prior to launch. An hour seems like a long time, but to get all weapons checked and aligned and all the gremlins out of the system, oftentimes we needed the whole hour to troubleshoot. Taking off and landing on the boat and flying through sandstorms in combat was tough on the jets, and sometimes trying to get the planes cooperating felt like coaxing an ornery toddler. When flying on the boat, we all become grease monkeys in our own way, learning how to troubleshoot and help our maintainers fix our $80 million airplanes.

My mother sometimes tells a story about me when I was three years old. I wanted to take the training wheels off my bicycle. She told me no, but two hours later, caught me with a wrench, removing the wheels anyway. She watched me ride without training wheels until I crashed, but instead of asking her to put the wheels back on, I picked up the bike and tried again.

Now, twenty-three years later, I was sitting on top of a Super Hornet, a wrench in hand, twisting bolts, getting dirty, trying to take the training wheels off.

Once the plane was good to go, we went through a host of comms checks with the boat to test all of our encrypted systems and then waited for the catapult to send us rocketing from zero to 170 miles per hour in just two seconds. Once airborne, we joined our wingman, and together we checked all our bombs, sensors, and comms again, and as a flight of two, we proceeded north, feet dry over the mountains of Pakistan.

During the transit north into Afghanistan, we flew on the boulevard, a super-highway for military aircraft that borders Iran to the west and Pakistan to the east. On the boulevard, just like on a road, we followed a myriad of rules. If you passed, you passed on the left, etc. These regulations and different altitudes kept us deconflicted from the host of Air Force, Navy, and NATO jets that were headed up and down the airway. Like the trucking days of old, everyone communicated through various bands of radio and encrypted voice communications channels as we passed from one checkpoint to the next. While flying on the equivalent of the I-95 of the Middle East, we also practiced tactics and warmed up our brains and aircraft systems for the day, because as soon as we crossed the border into Afghanistan, it was game time.

Immediately when we got in-country, we talked to the tactical air traffic controllers who we called Pyramid, and they sent us directly to our tankers so we could fuel up from the transit. In-country, we got most of our gas from Air Force tankers, arcing in huge racetrack patterns miles above the rocky terrain. Once they received our bunno numbers (so they could bill the Navy for the gas), we plugged into their baskets. The most common and most dreaded tanker aircraft was the KC-135, better known as the iron maiden, the kind of plane Tyler was flying when it crashed en route to Afghanistan. The iron maidens were loathed by naval aviators because their baskets were hard, and were suspended by a short hose connected to a long iron boom, which made them extra hard to refuel from. I never minded the iron maiden because it always made me think of Tyler. It was rare that I got close enough to see the pilots in their massive jets, but sometimes I'd just hear their voices and imagine Tyler, still at the controls.

In the Air Force, just like the Navy, the junior officers are the workhorses. The people in the sky flying with us were generally young guys like us from home and countries like Australia, Britain, and France. We were all there doing a job and talking to each other, our conversation alternating between business and chitchat. We often asked each other about our respective hometowns, and what we were eating for lunch or dinner in the plane that day. The Air Force always had better food than the Navy. They also had bathrooms and coffee in their aircraft, but I still wouldn't have traded my sexy jet, with its cramped cockpit and broken AMXD, for their big fat buses with Hot Pockets and lavatories.

After fifteen minutes plugged into the tanker, we were complete and ready to go to work. Pyramid told us where to support the troops on the ground, whether they were British commandos, Spanish soldiers, US Marine Special Forces, or Army infantry units. When we were operating and providing air support for ground troops, as the WSO, I was the one constantly in communication with the joint terminal air controllers embedded with their units, looking for bad guys, providing overwatch protection for their convoys, and asking what they were eating for lunch. From the sky, I could see everything—giant, craggy mountains, extreme shifts in elevation, villages of mud and brick, American forward operating bases, and deep, smoky river valleys. The physical beauty clashed with a culture that seemed forever doomed to war.

Three hours into the flight, I typically lost feeling in my butt cheeks and my shoulders would have a dull ache penetrating through them into my neck. By hour five, my lumbar region screamed for mercy. The jet seats are intentionally hard so that in case we eject, there's no give, otherwise, it could snap our femurs upon ejecting. When we started the long transit home a half hour later, the fifty pounds of gear on my body in my flight suit, my G-suit, my harness, and my survival vest would start to torment me. I'd imagine myself straitjacketed in an asylum. Joint helmets, because they were so expensive, were just passed along from aviator to aviator, and so the hot spots made it feel like all my hair follicles were on fire from the pressure of my ill-fitting helmet. From this point in the trip onward, it was a battle of wills. By hour six, I usually had to pee so badly I wanted to chew through a leather strap, but I'd learned my lesson.

At six and a half hours, the discomfort was so intense, it was almost comical. But with the most difficult part of the flight complete and the plane on autopilot, I could relax a little and open a magazine—*Architectural Digest, Vanity Fair,* and yes, *People*—and snack on breakfast bars and dried fruit. At some point, I'd pass Taylor in her jet, on her way inbound to replace me. She'd likely be reading the magazines I'd finished on my last flight and enjoying a lollipop. She loved lollipops in the cockpit. We'd pass each other, each going about five hundred miles per hour, and my plane's touchscreen would signal a text. Yes, American girls and boys are even texting in F/A-18s over a secret, encrypted communication system.

Everybody sucks but us.

"What did she say?" Crocket asked, hearing my giggle.

"Nothing." I checked my flight schedule in my stack of papers on my kneeboard to make sure she wasn't flying in a section with the skipper or XO. (You didn't want to send any joking messages with "mom" or "dad" on board.)

Go get 'em. I'll have dinner waiting for you when you land.

CHAPTER TWENTY-SIX

★

February 2014; Virginia Beach, VA

As deployment approached and officers and enlisted prepared to head out to sea for nine months, we tried to pack in as much life as we could. For departing mothers and fathers, it was a priority to spend time with their children. I, on the other hand, was swamped with a myriad of personal matters—moving out of Ashley's beach house, figuring out where to store my car, what to do with my sailboat, setting my bills on auto pay. Those were your standard leaving-town-for-a-while type things. But there was also getting a lawyer to draft my will, my DNR, and having four different powers of attorney notarized. I knew that I wouldn't be able to access anything on the ship's sluggish Internet, so I created a binder for my mom including all usernames, passwords, banking and investment account statements, and life insurance information. She was in charge of anything and everything that might arise over my nine months at sea.

Returning home from travel began to feel empty. Arriving at the hangar I'd watched aviators run into the arms of their loved ones. I'd haul my bag alone to the parking lot, often in the rain, to find my Porsche Cayenne—yes, I upgraded. Toss the bags in. Hope that the car, after sitting for weeks, wouldn't need a jump. At home, I'd throw open the window to my living room to let out the stale air and open the fridge to find a single rotten head of lettuce, a month-old slice of dried-up pizza, two Stellas. I'd drink a beer by the washing machine rumbling with weeks of laundry and imagine a husband, maybe a kid, and a warm dinner. I understood the desire for family and longed for something, ideally someone, stable in my life.

★

I wasn't alone. During the period leading up to deployment, it's not uncommon to try to find a romantic connection. Friends were pairing off right and left. And I met a civilian I hoped might be a good match for me. He was tall and handsome, an outdoorsman who liked to use his hands. He came from a good family, was kind and funny, a welcome change from my brash and cocky coworkers. And most importantly he wasn't intimidated by me or my career. He managed a Lumber Liquidators, so I'll call him Lumberjack.

One day when he knew I'd be flying late, he came over, let himself in, and started prepping dinner. At work I was engaged in some basic fighter maneuvers, or BFM, over the Atlantic when my opponent made a mistake while doing a stunt called a blind-lead turn, and we nearly had a midair collision. I'd debriefed and decompressed as much as possible, but nearly colliding with a jet that was closing with us at faster than a thousand miles an hour left me a little twitchy when I arrived home.

Lumberjack, on the other hand, had a peaceful day. He'd gone for a run and spent the afternoon making nachos, sipping beer, and watching TV.

"Hey," he called when he heard me coming through the front door. "Wanna slice tomatoes for the salad?"

I picked up a cutting board, placed a tomato on it, and started dicing when he stopped me.

"You okay? Your hands are shaking." He took the sharp knife out of my hand and sat me down on the sofa. "What's wrong? Something happen today?" His big puppy eyes listened more intently than any man's ever had.

I took a deep breath. "Well, I had a near midair during my flight today."

I described the incident, noticing those same brown eyes starting to squint and glaze over slightly.

"Sorry, babe. Remind me, what's a midair?"

As soon as he said *babe* I realized I didn't like it.

Jesus, I thought. "Okay, a midair is . . ." And I explained in a little more detail, using my hands to demonstrate the maneuver showing him the other plane was only about fifty feet away.

He squinted dubiously. "Wait, babe, fifty feet. That's not that close."

"Actually, when you're not supposed to get within five hundred feet of the other jet, that's frighteningly close. We nearly crashed because the other guy did a stupid move and disregarded all safety rules."

He looked out the window to pick out something five hundred feet away. "Fifty feet." His mouth made a little downward dip. "That still seems pretty far. How big is the plane?"

And then I tried to explain, using basic math, that our wingspan was forty-four feet, so at the speed we were approaching the other jet at the merge, fifty feet was like a stray bullet missing your head but singeing your hair. I could see he wanted to believe me, but he also wanted to comfort me by assuring me the distance was greater than I thought, and that I was not actually in danger.

"Caroline," he said. "You were fine. You are fine." Lumberjack then told me his own close-call story about nearly crashing his car in a Colorado snow drift.

Once again I noticed my hands shaking, but this time from anger, not fear.

I cut him off. "Three of my friends and I were milliseconds from either atomizing each other or ejecting from our aircraft a hundred miles offshore and swimming in the ocean with a hundred and sixty million dollars' worth of gear going down with us . . ." I stopped, hearing myself, and realized I wasn't being fair. Lumberjack couldn't possibly understand. The closest he was ever going to come to dying in the workplace was getting hit by a rogue forklift or a pallet of two-by-fours. I knew I was being overly critical, that I should compartmentalize what happened, have dinner with the guy, enjoy a back massage, and chill out. Those were the things I should have done. I also knew our relationship, or any chance we could be together, was done.

"Sorry," I said. "I know you're going to hate me. But I need to just be alone."

★

Ending things with the Lumberjack before they really got started was the right decision, and to be honest, an easy call for me. As much as I wanted to squeeze in a meaningful relationship before deployment, it was too much to bring him onboard, so to speak, before nine months at sea. Which was well enough, I would be headed to war clean. No guy at home, nothing to distract me. Able to focus on work and the business of combat. That was my mature self talking, the naval officer rationalizing the failures of my love life.

I'd given myself over to that inevitability when a few days later I was literally packing for the boat and my phone buzzed with a text.

Hey, buddy . . . can you talk?

Between the dangers of my job, the deaths of my two friends, and the challenges of my squadron, I was quite vulnerable. Ever since we'd broken up, the Minotaur had tried a number of ways to get me back, but for the most part, I remained committed to staying apart. Though with his crash, it seemed like something deep in him had changed and opened him up. Since I'd just ended the thing with Lumberjack, I didn't necessarily want to get back together with Minotaur. But I really wanted to talk to him.

He called that afternoon, and it was clear that he was single again, too. "I've done some soul-searching, Caroline. I've never really been fully there. I've made mistakes. Mistakes I wish I could change. And one of my biggest mistakes was not fully stepping up to the plate . . . with us."

"I wish you had, too," I told him. "But isn't that water under the bridge now?"

"Maybe not. Listen, I'm thinking about coming up to Virginia Beach for the weekend. Can I please see you? I have something to tell you." As he said this, I could hear the drone of his truck on the highway and knew he was already en route.

"Sure," I said. "But we're not sleeping together. And you have to do something nice. I need a little pampering."

My body was in rough shape. The strains of flying had taken a toll on my twenty-six-year-old body, my back especially. My spine and pelvis bore the brunt of the heavy gear and the poor ergonomics of the ejection seat, and while biweekly massages helped, I'd still lie awake at night, alternating ice and heat so I could walk without a limp the next morning.

Though a certified meathead, Minotaur had grown up with two sisters and understood women, especially how to take care of them. We went together—yes, this big, rough-and-tumble Marine—took me to a spa for massages side by side. Since his back was in rough shape from the crash, it turns out the princess treatment was exactly what we both needed.

After a morning spent getting pampered, he took me to Waterman's, one of my favorite places on Virginia Beach, where we sat outside and he ordered us fresh shrimp and orange crushes—a Virginia Beach signature cocktail made with crushed ice, Sprite, freshly squeezed oranges, and Grand Marnier. The day had been lovely, but there was one thing bugging me, or I should say, distracting me.

The Minotaur had decided to grow a mustache for no-shave November. But it was now well into the new year, and a nasty, patchy, scratchy caterpillar still hovered on his upper lip. Handsome as he was, it was hard to take my Channing Tatum lookalike seriously with a 70s-style porn 'stache. "You know," I told him, "that mustache makes my decision not to sleep with you a lot easier."

"Ha," he laughed. "It's part of my new look."

Okay, I wanted to say. *You think I'm kidding but I'm not.*

There are a lot of guys who will spring a ring on a girl before deployment. Many will profess their love, doubling down on commitment before separating for nine months. I wasn't sure if he was completely there or if I was ready for that kind of drama, but relaxed from the massages and a couple of orange crushes, I was ready to hear him out.

"So what's up?" I said feeling a little light-headed. "What did you have to come all the way here from North Carolina to say?"

"So . . ." He signaled the waiter for another round of drinks. "You excited to finally join the fight? You've worked your butt off. It's awesome you're finally going to fly in combat and do the things you've trained to do."

"Yeah, it'll definitely be an experience. I mean, you've already been there, done that, so I don't know why you think my first time will be any different from yours was. I mean, other than the mishap. Hopefully that doesn't happen. I don't know if I could fend off the Taliban like you did," I said, feeling myself start to ramble.

"You'll be fine. You're capable and good at what you do. Deploying won't be hard for you."

"I appreciate the vote of confidence, but I'm not worried about my skills in the plane," I said, playing with my straw in my frosty drink. "I'm worried about nine months of prison food and disgusting living conditions and squadron bros who don't want me there . . . whatever, though. I'll be fine."

"I definitely want to be there for you . . ." he said, reaching a hand across the table. *Here it comes,* I thought. *He's going to say we should get back together.*

". . . So I volunteered for a set of orders to Afghanistan. I'm leaving in a month and a half," he said, smiling at his manly gesture. "I'll be there when you are."

"Are you out of your mind?" My voice was loud enough to turn heads at the tables around us. "No one volunteers for deployment. The military—especially the Navy and Marine Corps—will get their money out of you and deploy you whenever they can. There's no need to raise your hand. You just got back." I stared deep into his eyes that for once didn't look a thousand yards away. "Why in the hell would you do that?" I dropped my voice to a whisper.

"It's not something I *would* do. I *did* do. I'm going because I want to be close. I want to make sure you're okay."

"But I'm going to be living on a boat and flying in my jet high over the country. You're not going to be able to take care of me while you're living on the ground in Kandahar or flying your helicopter."

"Yeah," he said, "but I figured I'd be in the CAS stack with you. If I can't be there on the boat with you, at least I'm not sitting here like a pussy while you're seven thousand miles from me, in harm's way."

The close air support stack, or CAS, is the tiered stack of aircraft that flies above the ground troops in combat, waiting to support the fight when called on. Usually in the CAS stack you have the helicopters flying down low, the unmanned aerial vehicles, or UAVs, conducting surveillance in the middle, and the jets high above. Everyone is layered on their own altitudes, a thousand feet apart, circling like vultures, hunting the enemy, ready to strike when called.

Did he jostle his brain in the crash? Where's Ashton Kutcher and the hidden camera, 'cause I know I'm being punked.

"So let me get this straight, you've offered to go back overseas so you can fly your helicopter in between heavily armed Taliban and Al-Qaeda fighters . . . and me?"

"I want to help protect you."

"You won't go to Phoenix, Arizona, to be with my family for Christmas, but you'll fly to Afghanistan?" I tried not to laugh.

He nodded, eyes locked on me.

"Okay, do what you want, but don't think this changes things," I said. "And unless you get rid of that awful mustache, I'm still not sleeping with you."

CHAPTER TWENTY-SEVEN

★

June 13, 2014; Embarked USS *George H.W. Bush*, Gulf of Oman

A civilian friend once asked me, "What's the coolest thing you can do in a plane?"

My honest answer—everything.

Every day, every minute, every millisecond in the jet was thrilling. I'd spent most of my twenties training and studying to be in that cockpit. I'd traded birthdays, weddings, and holidays to live like a vagabond and struggle to maintain any semblance of a normal relationship. But it was well worth it.

On my first combat flight into Afghanistan with Waldo, another one of my favorite pilots, I got to witness firsthand the awesome power even the sight and sound the F/A-18 has. We were supporting British commandos at a forward operating base. They were conducting surveillance on the Taliban when our radios crackled with a call about troops in contact, meaning our guys were engaged with the enemy—or more specifically, a group of Navy SEALs, one of which had the call sign Brutal, was in a firefight with the Taliban.

Our SEALs were posted up on a hilltop and the Taliban

were in an orchard. "Standby for a nine-line for an immediate show of presence." All right, it's game time.

We dropped down within earshot so the Taliban knew fighter jets were overhead. This didn't deter them; they kept engaging the SEALs. So we escalated our show of presence to a show of force. As soon as the order was passed, our planes dropped thousands of feet in altitude in seconds. Skimming through the mountains at two hundred feet off the farm fields, we raced toward the Taliban's location, executing a sneak pass at 500-plus knots. We swooped over Taliban terrorists like birds of prey dropping in for the kill. As per our briefed game plan, our jet and our wingman's jet came in from different angles so the enemy wouldn't know where we were coming from, so we could minimize our chances of being shot down.

As we screamed overhead, engines raging, we could see the bad guys plaster themselves to the ground covering their heads, some even curling up into the fetal position, thinking that one thing: If an F/A-18 is that close, it comes with thunder and lightning. They expected that, following the tremors and roar of our jet blast, they would feel their bodies being blown apart by our cannon fire or our bombing systems. No doubt there were some soiled pants and many bruised egos when they realized we were just sending a message—a message that was received. *You are messing with the best, if you keep this up you will lay burning in this field.*

The Taliban immediately ceased fire and the SEAL with the call sign Brutal and the rest of his team were able to get out of Dodge. We wanted to circle back and employ our weapons, but in this case, our ground commander's intent was met. We neutralized the Taliban without expending any ordnance, just using thirty-four thousand pounds of thrust and eighty decibels of the sound of freedom.

★

Supporting ground troops and working with other forces in the close air support stack reminded me of the Minotaur.

"Sometimes I think I'd get more news from a messenger pigeon," I joked to Taylor one afternoon. The Minotaur's emails had been slower and slower in coming. It bothered me, but the Band-Aid had been on and off so many times that ripping it off this time had lost the sting.

Looking out the canopy of my Super Hornet at the towering, snow-covered peaks of Afghanistan, I knew the sacrifices and difficulties with the squadron were worth it. I watched the attack helicopters in the valleys below me, and wondered if Minotaur was in one of them. Every time I got into the plane, I couldn't help but listen for his voice over the radio, even though I knew Marine helicopter pilots didn't talk to many people in-country. They flew too low, and if they did talk, they'd use different radio frequencies than jets. We flew high to stay safe, and they flew low to do the same, constantly hiding behind mountains, hunting the enemy just above the treetops. Just like in our relationship, I strained to hear the transmissions with so much altitude between us.

★

In Afghanistan, even though things started off slow, our mission ramped up quickly. After our arrival in late March, the first round of elections took place. We supported President Obama's request to provide security for the Afghani polling stations on April 5, 2014, and then again we were called in to support the second round of elections on June 14.

The first round of elections was rather uneventful, so as the second round approached, I wasn't nervous about flying in-country. Actually, I was feeling grumpy about flying into Afghanistan that day, because we'd shifted our schedule and I'd had to get up at two a.m. for my flight. I was flying with my skipper, Coma, and after the brief, we went down to breakfast and chose a table in front of one of the big screens. It was a luxury to watch TV over breakfast, because normally in the wardroom, we weren't allowed to have TVs on at mealtime.

That morning, Al Jazeera was doing a special on the Taliban in Afghanistan. They had reporters embedded within a small gang of the terrorists, and the bad guys were showing the reporters how to shoot RPGs and machine guns at the American base nearby. As the Taliban was terrorizing the base, a helicopter launched from the airfield. Initially, they got a little scared, until they saw it was a transportation helo, so they kept lobbing shots over the wall. Next an attack helicopter took off, and the Taliban got a little scared again, taking cover under a carport while still sneaking shots at the Americans. Finally, a fighter jet dropped down in a show of presence, dipping low enough that the Taliban could see and hear it overhead. At this, the group freaked out, yelling that the fighters had arrived and they must take cover. They frantically disbursed and scrambled toward the building like rats just before the news cut to a commercial.

I looked around at my four squadronmates who were mesmerized by what we'd just seen. We were stunned, staring at the TV, but then we unfroze and finished our breakfast in silence. We had seen what a show of presence from the sky could do and now we were getting the ground perspective.

"Whelp," Coma, our skipper, said, bringing us back to the moment. "In my eighteen years in the Navy, that was the most fucking surreal thing I've ever seen, especially before strapping into my jet to go fight those assholes. Guess it proves the work we're about to do actually matters. Let's go get 'em, boys."

"And girls," I said with a wink.

Once we arrived on station in Afghanistan, we were on high

alert. Just like we did for the first elections two months prior, we circled above the polling stations, watching for any nefarious activity on the ground and providing jet noise to protect the polling stations. As was evident by the morning's news, the bad guys were still there, lying in wait, biding their time until we let our guard down so they could assault the Afghani people with violence. We were ready for a long, stressful day when I heard a strange call from Pyramid.

"Hellcat 21, Pyramid. I've just received a call, you need to RTB USS Ship at this time."

I did a double take. *Can this be right?* Pyramid is saying that our section, Hellcat 21, needs to turn around and head back to the *Bush*.

"Pyramid, Hellcat 21 negative," I replied. "We have higher-level tasking and must remain on station," I said, basically telling the Air Force–enlisted air controller, located somewhere on the ground in Afghanistan, that we weren't going anywhere. We were there on orders from POTUS as watchdogs for the elections. In my mind, that was pretty high-level stuff, and somehow the guy working the radio for Pyramid didn't understand.

"Hellcat 21, Pyramid. USS Ship Actual says you will RTB this time. Proceed to your tanker at Gallop. Angels 25."

This time Pyramid's message was totally different. He said that the actual ship's captain, one of my many bosses, said we needed to get back to the boat pronto. And when the captain tells you to get your ass back, you move.

"Hellcat 21, copy all," I radioed back. "Proceeding Gallop at Angels 25." We headed to meet the tanker at his refueling track, called Gallop, at an altitude of twenty-five thousand feet.

Something isn't right, I thought, my mind totally blown.

"Skipper, has this happened to you before? I've never heard of them recalling anyone back to the boat so suddenly."

"Nope. Never have I heard of them calling everyone back.

Something big is going on. Look at the display—" He hesitated. "All the Navy tracks are headed toward the boulevard."

After topping off with gas, we all raced toward Mother like a swarm of bees headed back to its hive. What I couldn't see, looking ahead to the Indian Ocean, was the track of the *Bush.* In fact, as we arrived at the coast, the carrier still wasn't showing up on my system, so I dialed in the ship's navigational transmitter. Mother was at least eighty miles west of where we expected her.

What the hell is going on? I dialed in the RADALT settings for landing. Why are we going west? The plane caught the three-wire, and as we screeched to a halt, it hit me.

"Skipper, I think *Bush* and her entire strike group are getting sent into Iraq."

CHAPTER TWENTY-EIGHT

★

June 2014; Embarked USS *George H.W. Bush*, Strait of Hormuz, North Arabian Gulf

The boat's four nuclear-powered turbines spooled up as the ship's company battened down the hatches. My jet wasn't even parked, and I could feel the *Bush*'s deck heave forward as she pushed west at full power. We headed for the Strait of Hormuz, the tight bottleneck between Iran and the United Arab Emirates on the way to Iraq. Traditionally, tensions are high during straits transits, and since the Iranians were aggressive about contesting each American passage through the important waterway, the boat needed to be postured for battle. With everyone on high alert, Admiral Bullet and his lawyers positioned themselves on the bridge, and at midnight we navigated through the precarious strait.

Initially during all of this, we thought our movements were clandestine. The ship's captain had set River City, a communications condition that halted all transmissions to and from the ship. Of course there were still secret and top-secret messages getting sent, but it wasn't until the next morning when River City was lifted that I got a backlog of emails from home, clueing me into what a big deal our change of plans had been back in the States. I checked the timestamp on an email from my mom, noting that before I'd even landed on the carrier from Afghanistan, she knew where we were headed.

Well, it looks like you're going to Iraq. Be safe. Love, Mom.

As I read her words, I pictured her at the kitchen table, sipping on her morning coffee, wiping the smudges from her iPad as her eyes focused and then widened at the news splashed on the front page. She had set her Google Alerts for news about the *Bush* and Carrier Strike Group Two, so she knew the second it was public information we were turning west. In fact, given leaks, she might have learned sooner. Though I heard every news channel interrupted their normal broadcast to break the news that the *Bush* was on her way to Iraq. This was big news, as it meant President Obama was likely sending Americans back to a country he'd pulled out of, a country that was descending into chaos. This also wouldn't be the last time information about our mission would be shared real time with the public via Washington.

So much for the element of surprise, I thought, though in truth, it's hard to miss a carrier strike group when one is headed your way. And we still had the surprise of what our jets could do; that would be something the young ISIS terrorists had yet to experience.

When we woke at daybreak, we were at the northern tip of the Persian Gulf. Our operating area was known as the North Arabian Gulf, or NAG; this sea space would be the boat's home turf for the rest of deployment. Floating around the Gulf and flying into Iraq had decidedly not been part of our original deployment plans. Sure, we knew there was a possibility we could fly in the Gulf to project our power to Iran when they got edgy, but we didn't expect to be based in the Gulf or fly into combat anywhere but Afghanistan. As naval aviators, we were accustomed to operating forward and had the ability to be anywhere and launch into any country at a moment's notice. Now that we got the call, it was a race to get ready.

The US military and our tactical aircraft hadn't been in Iraq since the *Bush*'s last deployment three years prior in 2011. For all JOs on the boat, operating in Iraq would be a completely new experience. This provided the air wing with an entirely new set of challenges. All the charts, jet loads (computer files we load into the jet for every flight), radio frequencies, procedures, and kneeboard cards were totally different for Iraq than Afghanistan. Oh, and there wasn't a rulebook for what to do in Iraq. The pages and pages of ROE that we'd memorized were specific to Afghanistan.

While young junior officers scrambled to prepare for war in a new country, hammering on laptops, writing code, inputting data, and creating the smart packs with flight rules for how we would even enter or exit Iraq, the senior officers who'd flown in Iraq in the past grabbed a pot of boat joe and met in the admiral's conference room to brainstorm their tactics and approach. They knew the lay of the land and air. We didn't, so we had to make up for that. During one of our transit briefings on the Middle East, we'd briefly touched on Syria and alluded to problems going on in Iraq since the US pulled out in 2011, but in early 2014, we really didn't know the full extent of who or what ISIS was.

Our Internet returned the morning we arrived in the Gulf. I pulled up as many articles on Iraq and ISIS as bandwidth would allow. From the bits and pieces I gathered from the web and the intelligence officers who were pulling information from the secret network, I knew this much: The situation was beyond dire. Today we are all too familiar with the savagery of ISIS, the beheadings, the child soldiers, the massacres, the insanity. Today, at least for me, I have seen and read so much that I am inured to the barbarity. But then, when I first began to read about ISIS and what they were doing, I couldn't believe that a group so patently evil could exist. Every word I read, every image I saw stoked a fire inside of me that grew hotter and hotter, yet my exterior was cool. Game time was approaching. And I knew one fact with absolute certainty. My life and the last seven years of relentless training were all leading up to this moment. It was not only our purpose but our destiny to stop or slow down the tidal wave of murder sweeping over Iraq.

While I read silently, trying to understand the enemy we were up against, I noticed some of my roommates had a different reaction to the news.

"Oh my God, do you think this will extend our deployment?" One of them, Kelly, was visibly upset, shaking as she spoke. "I can't believe they canceled port, my husband is supposed to come see me! I haven't seen him in five months!"

After supporting the second elections in Afghanistan, we had planned to pull into port in Dubai for four days of rest and relaxation. Spouses, significant others, and kids had been counting the days until they could meet their loved ones for shore leave which, of course, was canceled when President Obama ordered us to go to Iraq. Families who'd not seen each other for months, who'd planned detailed vacations and spent family savings, were canceling tickets and hotel rooms that they were supposed to use in a few days' time.

Missing a port call was stressful to be sure. After almost half a year on deployment, we were all growing weary of boat life. I was just as frustrated as anyone, but the shift in our deployment was pivotal. We were needed to protect the civilians in Iraq, and for that, Dubai could wait. Sure, I would miss the suite Taylor and I had booked and getting a tan at the pool, but at least I'd save money and calories for the next time we pulled into the UAE. How's that for putting a positive spin on things?

Kidding aside, how could I be upset about missing a short port call when there was an insidious cancer growing inside Iraq that needed to be removed? Kelly, caught in the moment, didn't see it that way.

"We're going to lose so much money . . . shit! Our hotel is nonrefundable and this was the only time during deployment he could visit." She paced and swore some more.

Another one of my roommates tried to calm her. "Honey, I'm sure it'll be okay. Your husband will under—"

Kelly cut her off. "They can't do this to us! Do you think ISIS or ISIL or whatever they're called will be shooting at us?" she continued.

My nerves, already pulled taut, began to fray. The normally intelligent and logical Kelly was acting like Chicken Little, squawking about the sky falling. Talking in circles, winding herself up, she spiraled deeper into her fears and resentment. I couldn't sit on the sidelines any longer.

Turning around in my metal desk chair, I calmly said, "Girls, it's going to be fine. We're going to do our jobs. The mission we've been trained to do, and it's going to be exactly like we've practiced in training over the past four years."

"But we weren't trained on Iraq!"

"But you've been trained to adapt," I said, "to apply your skills and knowledge to a new situation. We've got to all calm down, remember that, and focus."

In her angst, Kelly couldn't acknowledge these truths. Finally, my other roommate calmed her down by embracing her in a deep, heartfelt hug.

As peace returned to the Sharktank, I turned my chair back to my laptop, slipped my headphones on, and cranked up my music. I needed to think and process. When faced with the unknown, or when shit hits the fan, a veil of calm descends over me. I don't freak out, I focus. With new information flooding in, I withdrew into myself so I could understand the problem and then methodically work through my role as we planned our attack on the human cancer known as ISIS.

Kelly was not the only officer freaking out. A lot of the JOs liked to stay between the lines, and now that the lines were changing, or in some cases even disappearing, the officers were pushed far outside of their comfort zones. It wasn't danger that concerned them. They were completely okay with war or risk and would gladly walk into a buzzsaw, so long as it was the buzzsaw they'd been prepared for. Now that the US government had put us in a different fight, they struggled and I thought about Val's motto: adapt and overcome. We all needed to draw from her inspiration.

But for the most part, the jet aviators did not freak out. For us, a pivot from the familiarity of Afghanistan to the uncertainty of Iraq was exciting. It was what we lived for and also what makes naval aviation so special. Whenever there's a crisis anywhere in the world, the first thing the president of the United States will ask is, "Where are my aircraft carriers?" Launching from a carrier can get US forces anywhere in the world within twenty-four hours. The Air Force and the other services, by nature, are a little slower. When the call came for the United States to go into Iraq, the Air Force and Marine Corps required clearance from their country of origin—even though they were taking off from a US base—to enter the new country. This involved convincing foreign heads of state, political considerations, and belabored bureaucratic processes—all causing delays.

The Navy, on the other hand, doesn't ask for permission. Our ships, and the enormous carrier beneath us, are all sovereign territory. We might have been parked off the coast of Iran, but thanks to our floating runway, we were in America, free to take off and fly wherever our president wanted us to go.

Twenty-four hours after receiving the call, our F/A-18s tore off the *Bush*'s flight deck, headed west into Iraq. That first day, only the senior-most aviators flew—CAG, Skipper Coma, and the department heads—and all had operated in Iraq before the 2011 pullout. On day two, still only senior aviators could fly in-country, but since I was one of the more senior JOs, I was allowed to support the planes going "over the beach." I flew with Lorde who was very senior and wanted as desperately as I did to get in on the action, but due to his country's rules and regulations, he wasn't allowed to enter into Iraq yet.

We launched off in a Super Hornet configured as a tanker with twenty-seven thousand pounds of gas onboard. Our job was to prowl back and forth along the border loaded with extra fuel in case any of our jets in Iraq needed gas for the trip home.

During the early days flying in Iraq, a cowboy culture existed downrange. It felt like the wild west. We knew so little about what was going on in Iraq that for every mission, we flew by the seat of our pants adjusting to an incredibly fluid situation. Since 2011, normal life had returned to Iraq. Unlike Afghanistan, which really had no civilian air traffic control, Iraq had civilian ATC directing commercial flights all over the country.

The abundant civilian air traffic flying around the same airspace with us made the first conversations between our combat-hardened aviators and ATC a little awkward. While Lorde and I played backup for the A-team, I eavesdropped on one of the senior WSOs conversing with air traffic control on the radio. The Iraqi air traffic controller happened to be a Middle Eastern woman with a heavy accent, wondering what in the hell our aircraft was doing in her space.

"Aircraft squawking 7203 in the vicinity of Basra International Airport," she said. "Baghdad Center, identify yourself." Translation: *I'm seeing an inbound aircraft on my screen I didn't expect, in the vicinity of Basra Airport in southern Iraq, and I want to know who you are because I don't recognize the code you're squawking.*

The WSO responded, "Baghdad Center, Hellcat 21, flight of two coalition aircraft, proceeding inbound, flight level 2-1-0. Prime Minister al-Maliki invited us." Translation: *Hi, ATC, my call sign is Hellcat 21, I'm in section with another coalition plane, but I'm sure you can tell by my accent I'm American. Anyway, we're flying into your country at twenty-one thousand feet to see what's going on. Oh, and don't give me any shit because your president wants us here.*

The Iraqi ATC responded again. "Hellcat 21, Baghdad copies all. Can you say type aircraft?" Translation: *Gotcha, I'm guessing you're here to help . . . but I really want to know if you're a fighter jet or a big fat spy plane.*

The WSO responded again, "Hellcat 21, reconnaissance aircraft." Translation: *Next question.*

Iraqi ATC then said, "Hellcat 21, roger, say ordnance onboard." Translation: *Yeah, right, I hear you speaking through your oxygen mask that sounds like you're in a fighter jet, but you said you're a spy plane. Whatever, I really just want to know if you have bombs and missiles you're going to use against the ISIS bastards who are encroaching on my home in Baghdad.*

"Hellcat 21, negative." Translation: *Cough . . . I'm not telling you . . . cough . . .*

The truth was our planes were always loaded to the teeth, ready to atomize any ISIS we found, but ATC, as a civilian controller, didn't need to know that. Still, she had a tough job. All day long, she had to deal with our jets perched over Iraqi hot spots, circling in unpredictable patterns, complicating her task of vectoring commercial airliners in to land at Baghdad International Airport, or BIAP. Also in the beginning, all of the normal Middle Eastern air traffic was still transiting through

Iraq to get to their destinations in the Gulf, so in addition to the air traffic controller's normally busy job deconflicting civilian air traffic, throwing in a bunch of US Navy aircraft added to her workload. I guess you could say that while ISIS was ruining her country, we were ruining her day.

★

As aircrew from the *Bush* successfully established procedures for entering and exiting Iraq, our Navy E-2s, the Three Bears included, did a great job with their new role controlling all the aircraft coming in and out of Iraq. Once we settled into a routine and got more confident flying, it was time to analyze what was really happening on the ground—where President Obama promised the American public we wouldn't put our boots. So initially, we kept our boots in the sky.

And starting with our initial strikes into Syria in October 2014, sometimes our "air" footprint wandered out of Iraq and into Syria, which in the beginning of our time in Iraq was especially dangerous. At that time, no high-value assets were allowed into Syria. So when we flew over the border, as fighters, we flew alone. We had no comms with our support aircraft, no tankers to give us gas in case we ran low, no boots on the ground, no one there to help us in case the unthinkable happened. And all the while we flew in Syria we had Turkish patriot missile batteries and F-16s locking us up with weapons solutions or bull's-eyes on our foreheads.

While in Iraq and Syria we performed reconnaissance and intelligence gathering in a few different ways, and each day we were assigned different points of interest based on our intelligence team's suspicion of ISIS activity. Our jets then flew into combat in three waves, each with a list of forty locations to search, like in a giant scavenger hunt. We would systematically fly to each location and scan the vicinity. When I was the lead

I always divided our sensor tasking so that the lead aircraft would look from the waypoint north and our wingmen would search south of the location. If we saw anything that seemed sketchy, we turned our tapes on and recorded all of our screens and sensors to bring it back to the intelligence department on the boat for analysis. Some locations didn't show anything of interest, like vacant warehouses, but we learned to recognize when ISIS had moved into a compound. The surrounding neighborhood would be eerily quiet, almost abandoned, and our place of interest would look like a frat house surrounded by dirty pickup trucks, coming and going at all hours.

I was flying in the vicinity of Ramadi and I couldn't help reminiscing about Minotaur and the stories he had told me about the intense ground battle ten years ago to retake the city. It was easy for me to envision him below, a young man scared, about to lose a friend, but it was hard to imagine that ten years later we were back, fighting a different fight.

A little off the main drag we started dialing our sensors into known intelligence hotspots when we found gold. Sitting up at twenty-five thousand feet looking down on a compound was always interesting. The hair on the back of my neck stood up as I noticed that the entire neighborhood was abandoned; everyone else had left their homes except for one place. Our place of interest looked like a group of troublemakers having a house party. Trucks and vans lined the driveway. Guys were hopping out of cars with armfuls of stuff. The kind of thing you would see during a boys' weekend except instead of a thirty rack of Natural Ice and Ruffles chips, upon closer inspection they were carrying AK-47s and other tools of ISIS's trade—paraphernalia of death and destruction.

We watched for an hour correlating groups that were coming and going. Finally as we tracked one group of young ISIS fighters leaving the compound for another neighborhood, I couldn't help but think they looked like they were on a snack run. The dust from the dirt road had barely settled as the group parked their trucks and eight guys hopped out. The men were on the move. They swept up toward a private residence. We watched them kick in the door as they broke into a house. A few minutes later they emerged in the backyard where they made a run for it, clearing the fence and subsequently hopping courtyard walls and fences as they made their way from house to house.

Eventually the main group separated into two groups that wove their way through estates, back alleys, and groves of lush trees. We tracked the groups. While reporting the groups' movements to our JTAC on the ground, Crocket and I took the lead group to the north and our wingmen took the stragglers to the south. This was exciting work, tracking the enemy perched high up above earshot and out of sight. It was challenging, difficult work keeping track of people moving on foot through a shaded and formerly densely populated area.

Crocket and I had approval authority to employ weapons but we didn't want to wreak havoc on those beautiful homes, which were empty now but someday would hopefully fill back up when ISIS was out and the innocent Iraqi people could come home. We didn't want them to come home to piles of rubble.

Finally we tracked the menaces into a big grove of trees as they regrouped and took a break. Bingo. Crocket lined up to strafe them with cannon fire. We executed the roll-in, overbanking and thundering down the chute, the earth getting larger with each passing second. It was very dangerous to dive so low. We could take enemy fire at any moment. But hopefully they didn't see or hear us until it was too late and we had already beat them in the shootout.

Then milliseconds before we opened fire we got a call from the joint terminal air controller, "Abort, abort, abort."

We instantly pulled off target, but because we were so low

the ISIS fighters now heard us and scuttled like cockroaches for cover in the nearest compound. To this day, I don't know why the ground commander aborted us. But after they did there was no more time. We were now running low on gas and needed to turn over the airspace to the next wave of fighters.

★

One day when assigned to search Highway 1, a four-lane interstate that runs north–south connecting Baghdad, Iraq's capital city, to Mosul, the largest city in the north, we received intelligence reports that ISIS planned to overrun Baghdad within seventy-two hours. Anchoring our sensors in the northwest suburbs of Baghdad, we scanned the highway heading north and found the road itself was quiet, little to no traffic heading in or out of the city, which was unusual for such a major thoroughfare. The hair on the back of my neck bristled. There was a reason the Iraqis were avoiding Highway 1.

My pilot and I discovered the cause—the four-lane highway was speckled with disabled vehicles lining the pavement. Peering down on the scene from twenty thousand feet, it looked like a train of sugar ants marching along one by one had been frozen in place. Stepping my ATFLIR, our infrared sensor, into its maximum zoom setting, I took a closer look. Most of the vehicles had the windows blown out, and dark stains seeped out onto the concrete and sand. My stomach churned; as we continued north, the problem worsened.

"Let's go around," I told my pilot. With our recorders on this time, we circled back around and tried to count the minivans, station wagons, and small pickup trucks abandoned along the highway. My pilot counted the southbound lanes, and I tallied the north as our jet whizzed by.

"I stopped at the mid-hundreds," I said, my heavy sigh audible even through the mask. "What'd you get?"

"I'm at least that high," he grunted as the Gs set in and we turned around. "Maybe we should loop back around to get a better tally and good video? Maybe they'll send it back to DC?"

I knew what we were watching and recording needed to be seen back in Washington. We weren't counting cars. We were counting commuters, families, couples massacred by the cowardly ISIS bastards. The checkpoints and barricades across the highway were clearly marked, the distinct black flags with white ISIS logos whipping in the breeze. The casualties were innocent lives—children and adults whose only crime had been to be on the wrong road at the wrong time.

"Let's do it," I said bracing myself in the turn and recalibrating the ATFLIR. "Jesus, I hope this is good enough to get their attention. How much damage do these assholes have to do before we intervene?"

That afternoon along Highway 1 was just one of many I spent in disbelief, watching from my safe bubble as Iraqis were plagued by indescribable brutality—the same trucks tearing through towns, shooting AK-47s at homes and village storefronts, the black flags again flying, heralding more death. Other aviators from the *Bush* witnessed mass executions. Men, women, and children lined up in rows, gunned down. Their bodies dumped into mass graves with bulldozers. We watched ISIS stockpile IEDs, pompously riding around in their marked trucks with MANPADS, surface-to-air missiles, that they would use to shoot at our planes. ISIS continually crossed over the Iraq–Syria border freely, flipping the United States and Iraq a big, fat middle finger. ISIS knew we were circling overhead, loaded with weapons and observing atrocities, but they also knew President Obama had not yet authorized the use of lethal force, so we were forced to hold our weapons. All we could do was continue to collect intel to help build the case for bombing ISIS.

In addition to hunting, we collected intel by working with allies on the ground. Because POTUS said he wouldn't put troops on the ground, these allies were referred to as "advisors" to the Iraqi army. But don't be mistaken, American special operations forces, or SOF, were on the ground in Iraq from day one.

The SOF teams ranged from Army Green Berets, to Marine MARSOC units, to Army special operations helicopter detachments, to platoons of Navy SEALs. Though our government didn't officially acknowledge any "boots on the ground," we supported them almost daily, providing overwatch for their convoys of black Suburbans during their patrols and securing their compounds.

In the event they came under attack, we could provide defense via lethal force in self-defense of friendly troops without any higher authority intervening. Providing support overhead, we positioned our two jets in the vicinity of the ground troops, but a couple of miles up in the air. By maintaining this standoff distance, we stayed out of range of ISIS missiles and hid our noise and visual signature so we didn't broadcast the presence of US troops. We flew our planes in big circles, like wagon wheels stacked vertically, but always on opposite spokes so that we could keep our sensors on the bad guys without interrupting with our video feed.

One day we had taken a defensive sensor posture, meaning that as the lead jet, we searched out in front of the convoy, looking for anything coming up on the convoy's route that might slow them down, or worse, reveal ISIS up ahead lying in wait. My wingman kept track of the convoy, searching for anything on their sides or coming up from behind.

I heard one of the SEALs communicating with me on the ground and noted a familiar call sign. "Hellcat 23, this is Oxblood. Assume defensive posture." Oxblood.

"Oxblood, Hellcat 23. Sensor posture set," I said, letting them know we were ready to go. "We've scanned your posit.

You and your route are clean." We spoke in a sort of code on the radio to operators on the ground, like a foreign language.

"Hellcat, Oxblood copies. We're on the roll. Advise of anything that seems out of place," one of the SEALs said back.

"Hellcat," I said, letting him know that I understood.

In conversations with the troops on the ground, I tried to be as succinct as possible because, like us, they had two or more other radios they were monitoring and transmitting on at the same time. We always used radio discipline, never wanting to clobber or distract the SEALs with useless comms and chatter while they operated outside the wire.

A few minutes later, I found a hazard on their desired route. "Oxblood, Hellcat 23."

"Hellcat 23, go."

"Oxblood, I'm seeing a disabled vic five blocks ahead, just before your planned left turn. Recommend you turn at the next intersection to avoid the commotion and backup."

"Oxblood copies," he said, and I watched the lead vehicle turn left as I'd instructed and veer onto the highway for a couple of miles and then got off at their exit and was back in a congested neighborhood.

My wingman came over the radio. "Oxblood, Hellcat 24. Be advised there is a group of pax approaching your posit from the east. They are passing the covered fruit cart right now and look like they're on a mission. Looks like five military-aged males." *Definitely nefarious.*

I zoomed in and saw the men approaching. Just as the bad guys were about to round the corner and approach the last vehicle of the convoy, the SEALs accelerated, ditching them in a cloud of dust.

"Hellcat, thanks for the heads-up. Those guys were definitely packing heat," one of the SEALs said. We couldn't be sure the bad guys were ISIS, but we knew they were up to no good.

We kept circling, diverting the SEALs around a few more sticky situations, slowly drifting a few miles south. My wingman and I yo-yoed to and from the tanker, making sure that one of us stayed overhead, on station, to provide the convoy an eye in the sky at all times. As our time drew to an end we turned over to the next set of jets and checked off station.

"Hellcat, thanks for the cover. Stay safe up there," the SEAL said as we departed for the tanker. "And how the hell you guys land a jet on a boat is beyond me." We peeled off and I never learned who this faceless SEAL was.

★

Flying in Iraq fascinated me. After just a few flights in-country, the pattern of life became very familiar, even from twenty-four thousand feet up. I could look down at the neighborhoods we searched and visualize myself on the ground, placing imaginary faces with real voices. Even though I flew miles above, I felt like I could touch the fruit I saw in my video feed of the farmers' market, and I could smell the crowd around me. It was this ability to put myself mentally on the ground with the Iraqis that helped me feel when something just wasn't right.

Afghanistan was a totally different beast. Like a Smurf Village for terrorists. Looking down into the craggy, smoky mountains and swampy wadis, you knew (whether you could see them or not) the country teemed with bearded villains hiding in caves, cradling AK-47s and RPGs while drinking poppy tea laced with opium.

Iraq, on the other hand, was very populated, almost cosmopolitan in places. Flying in-country, we crossed the coastline, passing Iran to our right, as bleak as an abandoned ghost town, and beyond that, oil fields dotted with flames from the refineries. From the air, Baghdad reminded me a little of Washington, DC, a maze of concentric circles with the Potomac weaving through the city center. Iraq's capital has a similar layout with the Tigris flowing through the middle, and sitting smack dab in the heart of the city, beautiful high-rises, fancy government buildings, nice houses, shopping centers, suburbia, and, of course, slums. Again, kind of like DC. The landscape along the Tigris is lush and glows green in the morning light, and on the outskirts, there's rich farmland and plentiful crops.

One evening, as the sun set over the city, I watched a carnival in the dimming light, spinning swings and a giant rotating Ferris wheel churning slowly through the desert dusk. I could almost smell the shawarma, and hear the calliope whistling, the children laughing. Why did Baghdad feel so familiar? Was it because the Iraqis were like us? Was it the classes we'd been taking on the *Bush*? What made me think I knew them?

The SEAL team sent us encrypted photos on our secret network so we could see the compound they'd turned into their operations base. The purpose was to help us orient ourselves, to see where they were located and better understand the minimal infrastructure they were working with. I studied the rudimentary facilities, but like the other Jet Girls, I also couldn't help noticing the gorgeous, jacked guys standing in the middle of the photographs.

I found myself also intrigued with Mosul, but for different reasons. The city had been taken by ISIS and was finally reclaimed by the Iraqi government in July 2017. When looking down on its western bank, just left of the Tigris, there's a very large pool. Even from high in the atmosphere, I could tell it was beautiful, and after slewing my sensors on it during my first flight over the city, I noticed something unusual—large, Olympic rings outside the entrance. During late June, the temperature was 120–130 degrees in Iraq and the Gulf, so I'd fantasize about diving into the water and splashing around. I assumed local Iraqis would want to do the same, but every time I zoomed in on the deep pool, I never saw anyone in it. Several years later I would learn why. When ISIS took over Mosul, instead of using the pool to teach swimming lessons and encourage aspiring Olympic athletes, the jihadists turned the pool into a killing ground. Locking Iraqi civilians and prisoners of war in gigantic steel cages, they drowned people en masse. If I hadn't seen pictures of the barbaric acts in that exact pool, I wouldn't have believed humans were capable of such cruelty.

CHAPTER TWENTY-NINE

★

Summer 2014; Embarked USS *George H.W. Bush,* North Arabian Gulf

From May 28 until August 2, the *Bush* remained at sea. That's seventy-two days with no port calls. Normally, the Navy tried to make sure we pulled into port every month, but with the dire situation in Iraq, and Ramadan, which occupied almost the whole month of July, we stayed at sea. Ramadan, the sacred time of fasting and religious practice, is so important in the Middle East that the Navy tries to keep Sailors respectfully out of port and away from trouble during its observance.

One small respite during this period of time was a special care package that arrived on the boat for the Jet Girls. The care package was from Ashley, who had gone on deployment with her squadron ahead of us and had recently gotten back to the US. Since she'd been in the Gulf just a few months before, she knew exactly what a group of girls would want and need.

As soon as we got the box into the Sharktank we tore it open. Charcoal face masks, EOS lip balm, neon sports bras, dried apricots, snacks, glitter nail polish, our favorite magazines, lollipops for Taylor in the plane. In the care package she included a letter with all the gossip from Virginia Beach plus crucial tips like which salon to get our hair done at during our next port call in Dubai. My mother had sent plenty of care packages to me, all intimate, thoughtful, and loving. But nothing my mother could send could rival a care package sent by a Jet Girl who knew exactly what we needed.

But eventually no care package from mom, a Jet Girl, or Mrs. Santa Claus could give us what we needed. We needed to get off the boat.

Seventy-two days is a hell of a long time, especially in the summer heat in one of the hottest places on earth. To give you an idea of how long seventy-two days at sea is, the Navy, after day forty, entitled us to a beer day, which meant a day off from flying and a barbecue on the flight deck with a couple of warm cold ones. Since the consumption of alcohol is forbidden on US ships, when the Navy issues you beer at sea, you know it's getting bad.

At this point it was the little things that got me through—the rewarding feeling of crisply striking a day off the calendar. *One more day gone,* you could tell yourself and look at row after row of days completed. One of the very valuable lessons I learned during Plebe Summer at the Academy is that while the all-powerful Navy can have nearly complete control over your life, and seemingly have power over the sea and the air, the Navy can't stop time. No matter what challenge the Navy is putting you through, no matter how impossibly hard things get, time moves on and there is a limit to everything, even deployments.

Another help at this time was my mother. She would religiously email me, telling me everything that happened at home, including all those mundane and deeply boring-to-most details of life back in the States. During the summer she was at Grammy's cottage on Torch Lake, she told me about the temperature in the mornings, the wind conditions, the food she was cooking; she described a particularly bad yet funny-looking way my dad wiped out while waterskiing. There was no detail she left out and I could read her emails and for a moment escape the boat and waterski with my dad. In return I would tell her everything I could about the boat. The good and the bad. And as we neared seventy-two straight days at sea, something very creepy was happening.

One afternoon Carolyn and I were approached by an officer who was part of the ship's company, meaning he was not an aviator, but a member of the crew. "Girls," he said, looking right and left, sketchily checking his surroundings. "Follow me." He pulled us into a back office. "We need to talk." After much hemming and hawing he got to the point. "I've recently been informed that you've been running a brothel out of your room."

"What?!" we both exclaimed in unison.

But the officer—our peer—just stared.

"Wait," Carolyn said. "Is this a joke? Someone really said that?"

He nodded. "Yes. One of my Sailors. He heard it from multiple sources."

I chimed in. "Surely you nipped it in the bud and told him to stop gossiping about officers, right?" I'd known this officer from my days at the Academy. I went on. "You know me. You know I would never do that."

"Well," he said, fumbling. "Let me just tell you what I heard. One of my enlisted reported that he was in your hallway the other night and saw men coming in and out of your stateroom all night long, and they were half naked with only towels around their waists!"

"How did they know it was *our room*?" I asked.

"Well, you know, everyone knows about the Sharktank, and he said that the room had a 'Girls Only' sign on the door," he smirked. "Why would a dude's room have that?"

Carolyn and I both looked at each other. We wanted to

laugh, but we also wanted to punch the guy. Carolyn, who kicked ass so much harder than 99.99 percent of the guys I have ever met in the military, had also suffered harassment. She had grown up wanting to be a fighter pilot as long as she could remember—before such a thing was even possible for women. Her father was a Navy pilot and she'd been born with the dream. A dream she had fought for. She'd risen to the top of every group she'd ever participated in and she could fly circles around most of the single-seat dudes I knew, and that only spurred jealousy and resentment. Even her call sign, TATL, was an inside joke with cruelty in its heart. The guys who gave Carolyn the name TATL, which stood for The Accident That Lived, told her the nickname came from a biking accident she'd bounced back from, but behind her back they joked otherwise. With profound dignity, Carolyn ignored those jokes and even embraced her call sign. But now with this clipboard-carrying weasel openly questioning if she and I were operating a brothel in our stateroom, it was almost more than we could take.

Almost breathless with anger, she reined her emotions in and calmly asked, "Was the sign on the door pink with a Minnie Mouse logo?"

He squinted. "It was . . ."

"Was it a Minnie Mouse sign?" I repeated Carolyn's question.

"Well, yeah . . ."

"You know that was a prank, right? That room with the sign on it is a guy's room," I said. The officer turned bright red. Carolyn laughed, but I was fuming. What this meant was that an enlisted Sailor had been hiding outside what he wrongly assumed was the Sharktank at all hours of the night, in a hallway that was supposed to be restricted to officers only. The Minnie Mouse sign that read "Girls Only" was posted on our neighbor's—fellow male aviator's—door as a joke. The Sailor told his friends and his officer boss a ludicrous story and the officer blindly believed him.

"Did you even fact-check this kid before approaching us?" I asked.

"Uh, well . . ." he stuttered. "Um, I told the guy I would check it out."

I started to boil. "But instead of actually taking a half second to 'check it out,' you just accused us of running a sex shack? You're really that dense? There are only six fixed-wing female aviators on this five-thousand-person ship and you thought we would throw our careers to the wind so we could make some cash on the side?"

"Well, that's . . . uh . . . what . . ." The officer continued making petty excuses.

"Dude," I told him, "you're out of your lane. We are done here." Carolyn and I left him and went back to our stateroom feeling infuriated, let down, and a little unsafe knowing that Sailors who were not supposed to be in our restricted area had been lurking at all hours. Moreover, we longed for the reprieve of port to just get off the boat and escape the grind, if only for a few days.

★

September 5, 2014; Dubai, UAE

Dubai is Vegas served in a pot of couscous, the perfect place to blow off steam. Taylor, Carolyn, and I splurged on a club-level luxury suite in the Grosvenor House, a sleek, London-affiliated, four-and-a-half-star hotel on the beach. But before we even checked into our rooms, the first order of business was brunch.

In the Middle East, the weekends fall on Friday and Saturday, and in the more liberal locales like Bahrain, Muscat, and

Dubai, Friday brunch is the main event of the weekend. Up to this point in my life, I'd had plenty of epic brunches stateside, but the smorgasbord we devoured that day in Dubai is, to date, the best I've encountered. Thanks to a solid month of meticulous research and planning by the Jet Girls, about twenty of our squadronmates dined at the fanciest Brazilian steakhouse in Dubai with a waitstaff of five bringing us round after round of filet mignon, lamb, and chicken churrasco fresh off the grill, along with all the mimosas and cocktails we could drink.

On one of my trips to the ladies' room, I found myself with a group of women in abayas, the head-to-toe robes worn by local Muslim women in Dubai. I could only see their eyes, then seconds later, they pulled off their flowy robelike dresses and what I saw underneath made my heart stop.

"Gucci . . . Versace," I whispered to myself, commenting on the gorgeous clothes hidden under their robes. Looking perfectly chic, the women touched up their makeup and took selfies on their iPhones before draping their scarves and coverups back on and heading out. It was as if someone dropped a cloak over a cage of magnificent birds. I was mesmerized.

Stumbling out of brunch, we were all a hot mess. It was time for an afternoon siesta, though not all of us made it to bed. Later that night, I heard stories—one pilot passed out in a shopping cart, in his hotel room, on the twenty-sixth floor. Another passed out midsentence with his pants around his ankles while FaceTiming his wife. It was the perfect kickoff for a port call.

Not that our penchant for partying was something to brag about. While some may think fighter pilots are cocky playboys and party animals, the truth was, in stressful, rigid environments like ours, blowing off steam became part of our natural rhythm. It was necessary for our sanity. When aviators pulled into port, we let loose to unload the stresses we carried from combat, the squalid living conditions, the near misses, the months spent tirelessly working in the most austere environments. And though most of us wouldn't admit it, the things we'd seen and experienced in Iraq weighed heavily. Witnessing ISIS's barbarity and utter disregard for human life was burden enough, but what really bugged us was that we couldn't do anything to stop it.

President Obama was still of the mindset that ISIS was the responsibility of the Iraqi government. Militants from the Islamic State of Iraq and Syria were murdering people at such a shocking rate statisticians couldn't keep up, but we hadn't been authorized to use lethal force. We had trained for years to counter this kind of threat. We could have saved countless lives. But we were being told by our senior-most leaders, many from ivory towers seven thousand miles away, that the situation in Iraq wasn't bad enough yet. Our hands tied, we had to exercise self-control and patience. The VIPs and international media crews onboard the *Bush* recording our every move only intensified these frustrations.

So when we finally hit port in Dubai we spent four days in the swimming pool, drinking twenty-dollar beers and forty-dollar cucumber cocktails like we were at a two-dollar happy hour. It was as if we were getting back a little of the college life many of us had missed, hanging out with fifty to sixty of our favorite air wing friends and fellow aviators, and making trip after trip to the swim-up bar.

Several times during deployment, I'd heard that pulling into port in Dubai and Bahrain would involve lots of "man stew," but I never fully understood the term. Well, after seeing fifty guys in a pool, the name became clear and even clearer when I noticed that after nine hours, none of them had gotten out of the pool all day. I peered into the slightly murky lukewarm water.

"Taylor," I said, slowly turning to the Sharktank ladies, my nose crinkling. "We've all been drinking . . . all day. And none of these guys have gotten out to pee. Not once."

She shook her head, realizing the same thing at the same time.

"Man stew" was all she said.

CHAPTER THIRTY

★

August 6, 2014; Embarked USS *George H.W. Bush*, North Arabian Gulf

The day we pulled out of port, I had volunteered to stand duty as the squadron duty officer, or SDO. It was nothing new, in that I had stood it a handful of times since embarking on deployment, and it meant that my role was to run the operations and flights within our squadron. Essentially I would act as the commanding officer's direct representative and I would make all the decisions in his absence. I'd volunteered to stand SDO because typically the first day pulling out of port is light. Instead of spending eighteen or twenty hours on duty, the day would be short and easy because I didn't have to manage any combat flights.

After wrapping up my responsibilities that evening, I turned over the phones to one of our Sailors and was crawling into my rack for the night when there was an announcement over the ship's intercom.

"This is the TAO. Launch the alert 60 fighter. I say again. Launch the alert 60 fighter."

Just like the morning when Crocket and I intercepted the Iranian F-4, my heart immediately raced. I hopped out of bed and into my flight suit and ran to the Ready Room. I arrived, winded and shiny with sweat, and found my skipper with his kneeboard in hand, eyes wider and more alert than usual. A heightened sense of urgency had stepped up the already electric feeling on the boat.

This had to be the real deal. I bet someone had crossed a line, and it was time to bring the thunder. Someone very high up had ordered our jets into Iraq under the cover of darkness, and in the back of my mind, I suspected that ISIS was going to wake up to the shock and awe of our weapons on target.

I wasn't flying that night, but because I was SDO, I was in charge of preparing everything for the jets going out. I had three hours of work to do in less than thirty minutes so that my skipper and his WSO could lead their wingman into Iraq.

I went to work, again the stress bringing with it a veil of composure that settled over me and calmed me. Even though I was still a little worn out from Dubai, I successfully jumped through all the hoops to equip the men and their jets for the flight, with one exception. I hadn't been able to open our finicky gun safe to give them their pistols and extra magazines. Luckily our sister squadron loaned them their extras, which was a small reminder of the team effort required to get weapons downrange. By the skin of our teeth, two jets tore off the deck within the allotted hour.

The jets made it all the way to northern Iraq, ready to go, but never got the authorization to drop. As the duty officer, I stayed up all night long, watching the operation unfold in real time on my classified computer feeds. Finally around eight a.m., the aircraft landed back on board, bombs still intact. Exhausted from my twenty-four-hour day, and disappointed, I collapsed into my rack wondering when, if ever, we would be allowed to use force to save lives.

★

August 8, 2014; Embarked USS *George H.W. Bush*, North Arabian Gulf

Since I'd just been on duty for twenty-four hours, crew rest regulations mandated that I could not fly for eighteen hours after finishing my all-nighter. Still, after getting a good morning of sleep, I went up to the Ready Room to stand my short duty of the day in the tower, surprised to find the place was abuzz.

"Dude, today's the day! It's going to happen." I heard someone shouting from the horde of JOs hovering around the duty desk. "That artillery piece is definitely going to cross the line." Standing on tiptoe, I saw that the SDO had all his screens dialed into the live feeds from Iraq. Crocket read from a classified chat room, reminiscent of the days of AOL Instant Messenger, except all the transmissions were top secret. He rattled off the feeds like a seasoned NFL commentator, calling the play by play. "Hellcat 23 is overhead in the town of Irbil . . . an ISIS artillery piece is aimed at the city. Two F/A-18s are circling high above the target, standing by, with weapons solutions on the artillery, waiting for clearance from Washington, DC, to strike it . . ."

Who was flying in-country? I wondered.

I quickly checked the schedule and saw that Bobcot and one of our lieutenant commander WSOs were flying as Hellcat 23 for the day. Bobcot, one of my classmates from the Naval Academy, joined the squadron a few months before me. He was not your stereotypical fighter pilot. He *was*, however, your stereotypical nerd. At the Academy, he'd been a member of the Navy marching band, called the Drum and Bugle Corps; as a matter of fact he was the king of the D&B, aka, the drum major. While we weren't necessarily good friends in school, in the squadron, I'd come to know him as one of the good guys. He was the technically inclined JO in our squadron who fixed the computers that ran our planes, programming and coding them himself. He was brilliant, and flew with finesse. Later, he would go on to become one of the Navy's best test pilots, but in that moment, he was in the spotlight in Iraq.

Bobcot hadn't been handpicked to fly that day. He was just in the right place at the right time. Crocket read aloud from the chatroom feed, "ISIS is encroaching on Irbil." A call came in. President Obama finally authorized the use of lethal force.

We watched live from a drone feed on our secret network as a bomb, released from Bobcot's wing, tore across the sky and struck its target—an ISIS artillery piece.

When the bomb's blast cleared and the artillery piece was barely recognizable, looking like a pile of bent paperclips in a crater, cheers erupted across the Ready Room; our celebrations were as loud as a Latin American soccer stadium screaming, "Gooooooooal!"

Finally, we could fight.

★

In a surreal media-age twist, before Bobcot even landed back at the boat, the Pentagon had declassified the footage and pushed it to news stations across the world. News sources stopped their normal programming and began reporting on the first bomb to strike ISIS. We all knew what it meant. That first strike signaled a renewed commitment to the people of Iraq and to destroying ISIS. When Bobcot and his WSO stepped into the Ready Room, their first stop was to pause in front of the TV for a photo we took of the two aviators with their footage scrolling on CNN.

In that moment, I watched Bobcot with a mix of feelings: gratitude, excitement, a bit of awe, and even a little envy. We had all wanted to be the first to bring the fight to ISIS. Now, he had been the first man to drop bombs on ISIS. I wanted to be the first woman to do the same.

★

The following morning felt electric. As usual, I woke up early to get a cup of boat joe and a little alone time so I could read news and emails in peace. At the top of my inbox was an email from a familiar sender. I opened it and read the single line: *Caroline, go get some. -Me.*

The world was watching . . . and apparently that world still included the Minotaur, who had seemed to have dropped off the face of the earth in the past five months. He was back, at least, to wish me well. As for the rest of the world, the media now had finally shown in detail some of the carnage we'd been witnessing. Across the world, and especially in the US, people were rooting for the USS *George H.W. Bush* to relieve the citizens of Iraq from the horrors of ISIS.

Before my brief, I made my way down three decks to the vending machines to buy a cold bottle of water and a Gatorade. It was so hot that I had to replenish electrolytes so my body chemistry didn't get thrown off, making me sick in the jet. Hopping off the bottom rung of the ladder on my way downstairs, I took off at my normal speed-walking pace. I rounded the corner and almost slammed into Admiral Bullet. He was leading a small group of Arab men, a special group of VIPs from Saudi Arabia, on a tour of the carrier.

"Good morning, Dutch!" he said. "I'd like you to meet the executive assistant to the royal crown prince of Saudi Arabia." He gave me the man's name and then introduced the distinguished group. I smiled and shook hands, gulping air as I tried to catch my breath from racing down the ladder wells.

"Gentlemen," Bullet said to his contingent. "This is Dutch.

She's one of our F/A-18 weapon systems officers. That means she controls all the missiles and bombs in our fighter jets. Dutch, I'm just showing these gentlemen around. Care to join?"

Of course I did. I loved any chance to spend time with the admiral, and I had a knack for hosting visitors. My back was killing me. The nonstop flying had taken its toll, and I knew if I wanted, the admiral would change my flight schedule that day, giving me another day of rest and a chance to escort the royal contingent around the *Bush*. But I had a feeling . . .

"Sir," I looked him in the eye. "Normally I'd love to, but I'm getting my dream shot today. I get to fly in-country! I can't pass that up."

"Dutch, no explanation needed," he said, beaming. "Go get 'em!"

★

I flew north with Cardboard, Crocket's roommate, for almost two hours, all the way up to Mount Sinjar in northwest Iraq. Cardboard was one of the good guys: helpful, but a little bland and quiet, hence his call sign.

During our intel brief Cardboard and I received reports that ISIS had trapped thirty thousand to fifty thousand Yazidis atop a mountain. The Yazidis, one of the oldest ethnic minorities in Iraq, are believed to have descended from the Zoroastrians, and their traditions blend Christian and Islamic beliefs. The hardline ISIS militants consider Yazidis devil worshippers, and had already massacred some five thousand Yazidi men, women, and children.

Once we were finally on station talking to a JTAC, we searched the area looking for nefarious activity around the mountain. As we searched, Party 45, another section of jets (i.e., two jets—a lead and wing) from the *Bush,* joined us. All four jets hunted bad guys, dropping our sensors in different areas, searching for suspicious activity.

Cardboard and I were running low on fuel and about to head to the tanker when, like a hunting dog on point, one of the other jets alerted us to a potential threat. I dialed my sensors into the location. Looking outside and then down on my targeting screens, I realized my eyes weren't lying to me. I really was witnessing two tanks shooting fireballs from their main turret. On the screen we couldn't tell what they were aimed at, so we looked outside, seeing clearly that the explosions from the tanks were going into a village on the side of the mountain. We listened to Party 45, the lead jet of the section, call the situation into the JTAC, but as we did so, Cardboard and I had to go. Since all four of the jets on station would be refueling from the same tanker, we had to cycle through so that the next aircraft could follow us and the conga line didn't get backed up. We called the process yo-yoing because like a yo-yo, one jet always left and returned before the next plane could do the same. It ensured that all the aircraft could get gas, but at least one was over the action, keeping eyes on the enemy at all times.

As we flew toward the tanker, I dialed him up on our front radio but kept the tactical frequency up on the back radio. After my check-in to get gas, I passed the tanker frequency and comms over to Cardboard, so I could listen and track the situation at Mount Sinjar. I looked down at my kneeboard at my notes tracking the action and saw goosebumps on my wrist. Party 45 and our wingman had found two ISIS-controlled armored personnel carriers (basically tanks that also carried people), and an up-armored Humvee with a mounted rocket launcher. The men in these vehicles were lobbing grenades and rockets at a village on the north side of Mount Sinjar. Prior to departing for the tanker, Cardboard and I had seen the initial blasts with our eyes, but now the jets had gotten their sensors on the village so they could see people running from explosions—kids, old men, women—zigzagging every which way, down, across, and up the mountain.

Once we reached the tanker, Cardboard worked his ass off in the front cockpit, precisely maneuvering our jet to get the two-foot-long probe on the nose of our plane into the tanker's basket. This was no easy task, flying within fifty feet of the gigantic supertanker, bouncing around in turbulence. Cardboard got the probe in the basket and kept it there for the ten minutes we needed to top off with jet fuel. Since the tanker track was almost fifteen minutes away from the action and we couldn't have our sensors on while refueling, I had to keep track of the engagement with notes on my kneeboard and by marking up a chart I had of the area. It was a total mind game at this point. I was intent on mentally staying in the fight—listening to the radio, visualizing the movements on the ground, logging and inputting coordinates into our system—so that as soon as we got back overhead, we'd be ready to go.

After what seemed like the longest tanker evolution of my entire life, we were finished gassing up. I double checked my weapons and made sure they were ready. I reprogrammed my sensor so we were dialed in, should we be given the green light.

Check, check, check. Everything good to go.

I signed off with the tanker and switched our radios back to have the JTAC on primary frequency and our wingman on the secondary radio.

"Recommend gate," I calmly said to Cardboard, meaning that he should go max blast—as fast as the jet can go. I'd never recommended that before, but with every second that passed, I knew there were Yazidis on the ground getting shot up. We needed to be in the target area with our sensor and weapons ready.

"Sorry, Dutch," Cardboard said. "I won't gate. Don't want to waste the gas."

Even though he was an ever-boring pragmatist, Cardboard was right. In combat we flew at two speeds: max endurance, the slowest we could fly to preserve the most gas, or tactical airspeed, the speed we used to employ weapons and fight battles. Gas is life when flying fighter jets. We never have enough, and we have to be extremely careful how we use it. If we ran out over northern Iraq, at the very least, we'd be forced to visit the tanker, and at worst, we'd have to eject and risk capture by ISIS.

While he still didn't go full throttle, Cardboard bumped it up a bit, and I got the ATFLIR—the infrared targeting pod on the side of our jet—warmed up and calibrated for the environmentals in the area. As the day wore on and we started reaching the heat of the afternoon, we were constantly having to adjust our sensors and recalibrate the pods so we could see better during times of thermal crossover.

Still far away on the south side of the mountain, I could no longer see the tanks but I listened to the comms. We crested Mount Sinjar, and a few minutes later—*jackpot!* I spotted them again, preparing for their next volley of shots. I waited for a lull in the comms between the JTAC and the jets who were still on station and then called, "Hellcat 26, established Angels 19, captured," checking in with the JTAC and the special ops guy on the ground who was coordinating the planes in the area. I told him we rejoined the close air support stack, the cluster of Hornets and UAVs buzzing over the area, and that we had our sensors locked on the enemy. Along with the other jets overhead, I followed the tanks from where they'd been shooting to where they were parked on a highway, noting that along the way, they'd stopped and loaded the personnel carriers up at a small gas station, and we'd counted sixteen men between the three vehicles. The JTAC passed us a nine-line—the instructions for attacking the vehicles—so now, with the tanks retargeted, all we were waiting for to save the Yazidis was the authorization of lethal force.

It was only day two of kinetics in Iraq and little did I know, the weapons release authority was still held by the secretary of defense. So while we sat overhead, watching the bunch of jihadists massacring villagers, the JTAC's request was being routed all the way back to Washington, DC, and up to the honorable Chuck Hagel. I could imagine him, in a meeting in his office in Washington, receiving a phone call on his classified line from one of the admirals in the Pentagon, getting briefed, and authorizing the employment. President Obama had cleared Bobcot's strikes the day before, and now SECDEF was authorizing the use of lethal force.

As we waited, communications between the Special Forces JTAC and our jets became unreadable, but luckily there was a drone circling low over the target that relayed the radio calls for us. Finally, after what seemed like an eternity, the word came. Our lead jet, Hellcat 25, was on the tanker, and the two Party jets were getting desperately low on gas waiting for the employment. The JTAC ordered the Party flight to go refuel and authorized us, Hellcat 26, to employ.

"Hellcat 26, cleared hot, cleared hot."

Those were the magic words I'd been waiting for. Cardboard and I now had authorization to employ two GBU-54 laser-guided JDAM bombs. Game on!

Cardboard momentarily widened the circle we were flying until he reached the apex and banked the jet sharply around, pointing our nose at the target. My bombs dialed in, I gave my French braid one last tug for good luck before the acceleration pressed me back in my seat. I had already recorded the setup while we were overhead and was perfectly targeted on the two tanks as we pushed inbound. But then, both of the vehicles started to drive off. The laser-guided bombs could hit moving targets, but I couldn't get them both if they were too far apart.

Which tank do I hit first? Or should I drop a bomb in the middle of the two? I had milliseconds to choose. *Nope, definitely the lead vehicle.*

Perched in the back of a Super Hornet, racing across the blue sky at close to five hundred miles per hour, I watched the tanks move on the black-and-white screen in front of me. In my hands were controllers that looked like they belonged to a high-tech PlayStation. I guided the crosshairs on my screen with a tiny joystick controlled by my thumb, basically playing the most important video game of my life, where I had to keep my pipper—the dot at the center of the crosshairs—on the lead tank. Delicately moving the joystick, I noticed a small cloud of dust rising behind the tracks of the tank. The plane wobbled twice, and I felt each bomb release. I glanced up from the pipper to the impact countdown.

"Hellcat 26," I said over the radio. "Thirty seconds."

The five-hundred-pound bombs, packed with 480 pounds of TNT, were literally at my fingertips. I focused as hard as I could on my screen, making sure they struck the lead tank at just the right spot.

Boom boom.

The bombs impacted.

The lead vehicle was blown up entirely, and the second vehicle flipped, partially destroyed in the blast.

"Hellcat 26 impact," I said as the shock waves reverberated out and back from the point of impact.

"Hellcat 26, good hits! Good hits!" the JTAC relayed through the drone. Coming back a second later, he gave further instructions. "Hellcat 26 authorized immediate reattack, your target is the up-armored Humvee. Cleared hot!"

"Hellcat 26," I said calmly, acknowledging his call.

I was cleared to go after the technical vehicle that was armed with a rocket launcher, which was still smoking from its recent attacks. On my left screen, I stayed locked onto the first bomb site, stepping through different modes and zoom settings to ensure we got adequate proof of the attack to bring back for intel. On my weapons display to the right, I selected the laser-guided maverick, or LMAV. I knew it would take the missile's internal brains nearly thirty seconds just to spin up, so I did not hesitate.

The plane vibrated, the Gs slamming me back in my seat. Grinding across the sky, the plane let out a metallic growl as we arced around the circle, reminding me that while I worked the brains of the jet, Cardboard was flying it, setting us up for the attack. Noting our position in the overhead pattern, I moved the joystick to find our next target.

"Hellcat 26 is in, in 30," I said once the target was under the pipper, letting them know I'd employ the missile in thirty seconds. At the ranges we were employing the LMAV, we didn't need to calculate the time of flight because the missile traveled so fast. Once the missile left our wing, it would impact the target less than a few seconds later.

I felt a shiver. This was fast. Very fast. It would be nearly impossible for a single-seat jet to reattack a new target with live ordnance that fast. That's why two-seater F/A-18s are so deadly. It's the teamwork. We can go after these fuckers because while Cardboard bent the jet in the sky, I could target terrorists, ready the weapons, and guide them in for a kill shot.

I had the missile on and ready, the codes were set. Laser on, laser code set. The RADALT, the piece of equipment that kept us from flying into the ground, was set to the heads-up display and dialed in.

Check, check, check. I flew through the steps, keeping one eye locked on the FLIR (the targeting pod that would guide the missile) to ensure I was slewed, or locked, on the truck.

"Captured, checks complete," I said over the ICS, letting Cardboard know we were locked on the target and the systems were ready for weapons delivery.

The Gs intensified as Cardboard tightened up the circle and prepared for our roll-in. Tracking a truck while rolling into a dive in a jet was like playing Call of Duty on a roller coaster. I wrapped my pinky and ring fingers around the hand controllers to stabilize myself while my thumbs and trigger fingers targeted the vehicle. I braced myself for what came next. Cardboard unloaded the jet, overbanking 135 degrees, simultaneously rolling inverted and pointing our nose at the target. A millisecond later we plunged into a nosedive, our fighter hurtling toward the ground at more than four hundred miles per hour. I didn't look up from the targeting pod, knowing that if I did, I'd have a face full of ground in the windscreen. No sky. This maneuver defies logic and delicately dances with physics. There's math behind it, and you need to keep track of the numbers so you don't die, but in that moment, there was no time to think about dive angles and offset degrees. We had to make sure we got the missile on target.

A half second later, we rolled back upright. Sticking my tongue out, like I do when I concentrate, I strained to center the pipper on the truck and its rocket launcher, and pulled the diamond over the target, the whole time squeezing the left trigger as hard as I could, unconsciously white-knuckling the hand controller. In an instant, like a super-powered magnet, the laser-guided maverick locked on the target and the symbology changed.

"Good spot," I tell Cardboard. "LMAV locked!"

Whoosh. The LMAV roared, accelerating ahead of the plane, and I nearly jumped. Normally you don't hear noises outside the jet, but the LMAV, screaming as loud as a ten-foot bottle rocket in your ear, is hard to miss. You never get used to the sound. In such a steep dive, the cursor tried to rise up and off target, but I kept the crosshairs right on the truck's gas tank, ensuring extra explosive capacity so that we didn't just get the vehicle but the gun mounted in the bed, too. The plane moved again, the Gs crept up my body as Cardboard started to pull out of the dive. The pullout on an LMAV delivery needed to be gentle, otherwise the missile could lose the laser energy, thus going dumb and spiraling out of control.

"Don't mask, dooooon't mask," I cautioned Cardboard in a soothing, low voice.

Schwack! My entire display disappeared into a white and black mushroom cloud. I tried to zoom out and then realized I was already zoomed all the way out. Holding the cursor on target, I stole a quick glance outside at the real-life explosion ballooning up from the pile of wreckage. Smoke roiled black and red as gases combusted in the cloud.

Whoa, well that had the intended effect. I smiled.

As we pulled up to fly out of the missile's fragmentation envelope, Cardboard gently stood the jet on its left wing to keep our ATFLIR on the target. The job wasn't over yet; we had to record the entire explosion, getting it in all different modes of the sensor.

There was no way to sugarcoat the aftermath. Bodies spilled out of the Humvee, taking herky-jerky steps like zombies before collapsing dead on the desert sand. More mangled bodies scattered around the tanks, and still more inside, incinerated in the smoldering vehicles. I knew the dead were many, and only later, back on the boat, would I learn the official estimated count—sixteen. Sixteen enemy killed in forty-five of the most intense seconds of my life.

Our wingman polished off what remained of the second armored personnel carrier with another laser-guided bomb, and once we were sure there were no survivors left to terrorize the Yazidis, we headed back.

The cockpit was quiet, Cardboard and I, deep in thought, each knowing we needed to be buttoned up for the close scrutiny that would await us. Every move, every millisecond of our mission, would be scrutinized to ensure our employment was textbook.

Departing the tanker after gassing up, we tried to follow our lead in the descent, but something was wrong.

"Holy crap, Dutch," Cardboard said. He was normally a perfect formation flier but our plane kept sliding ahead, unable to slow down and fly in position off our lead.

"I can't slow down! What's going on?"

"Cardboard," I told him calmly. "They're still carrying bombs, but we're missing all of that drag."

"Oh . . . shit. Okay . . . ummm." Cardboard keyed his radio. "Hellcat 25, can you bump it up a little? We're at idle with the speed brakes out, and we can't slow down."

It was night when we got back. Cardboard caught the three-wire, and we screeched to a halt. As with Bobcot's flight, word of what happened had already reached the ship. When we finally parked, shut down, and climbed down the ladder, Cardboard and I were swarmed with cheers and high fives from the maintainers and ordnancemen. A female ordnanceman unclipped the dangling bomb-arming wires and handed them to me. "Keep it," she said.

"Thanks." I smiled. "And thank you for the awesome plane." I gave her a quick hug before tucking the wire into my helmet bag. Legs stiff from the flight, I stumbled to the tower so I could get grilled by Navy intel weenies.

"We did it," Cardboard said.

I nodded, then laid my shoulder into the gigantic door, which heaved open to reveal Admiral Bullet standing on the other side, grinning ear to ear. "You did it, Dutch. You got your dream shot and you knocked it out of the park!" He patted me on the back. "The admiral in charge of CENTCOM and I watched your bombs in live time. You shacked the target," he said with pride. "You really did it!"

CHAPTER THIRTY-ONE

★

Fall 2014; Embarked USS *George H.W. Bush*, North Arabian Gulf

As the dog days of summer droned on, the air wing settled into our new role employing ordnance on ISIS. The rest of our time in the North Arabian Gulf seemed to fly by. The Blacklions were doing well. In our mid-August change of command, the diaper-advocating Coma was replaced by Chick. We all knew Chick from the last eighteen months with him as our executive officer, and I don't think I was alone in my excitement for what I felt would be a positive leadership change.

I was thriving as well. Since the start of deployment, I'd flown more than forty flights in combat and had progressed from being a nugget, to a combat section lead. I'd been vested with the responsibility to lead two jets and three of my peers into Iraq and Syria to confront the enemy without any adult supervision. It was a big deal. No other squadrons had such junior JOs leading other JOs in-country, and it showed that my performance was well-respected.

When our squadron did its biannual junior officer job switch, I moved on from writing the flight schedule to supervising the maintainers and was loving my new role. As much as

I enjoyed being in the air and flying around at Mach 1, taking care of the people that kept our jets in working condition was surprisingly the most gratifying part of being in the squadron. It was the little things—ensuring they had clean bathrooms and showers, air-conditioning in the 120-plus degrees, or even just getting them watercoolers to cool off. I also worked hard to get them promotions or accepted into programs they applied for. It wasn't something I expected, and maybe I had Bullet's example to thank for it, but mentoring the enlisted who worked so hard on our jets made me feel like I was giving back.

Combat-seasoned and satisfied with my contributions on the ship, I was finally feeling like a member of the team. Of course I wasn't best buddies with all the guys in my squadron, but I had a solid group of guys I could trust, and a smaller, but even more loyal, group that I considered friends. Many of the guys I'd had conflicts with had moved on to their next jobs, and the remaining junior officers were all pretty copacetic. For the past seven months of the cruise, we'd all been so focused on the mission and working together as a team that there was no time for the petty, cliquey drama I'd experienced back home as the FNG. At sea, it was just about the work. No wives to stink-eye me from across the room; well, except for the group who stared at me from the three-foot poster that hung in our briefing space. There wasn't any division about who got invited out on the weekends because there was nowhere to go. We were all united in our efforts to survive the scary nights behind the boat and constant threats that were popping up in our missions. Sure, I spent lots of time holed up in the Sharktank, working on my master's instead of playing video games after hours with the bros, but for the most part, I'd found my place with them.

Before leaving the Gulf, we had one last Middle Eastern port call in Bahrain. And while it wasn't as intense as Dubai, it was our last time to cut loose and stock up on rugs and gold jewelry. By now, Taylor and I were the expert cruise directors and booked our squadron at an incredible resort that looked like it belonged in a coffee-table book you'd find in an Ace Hotel. Just like we'd done prior to deployment, we spent four days living like there was no tomorrow. We brewed a mean man stew at the swim-up bar, practiced our synchronized pool aerobatics, and cleared the local souk of their exquisite, handmade wares. Throughout the port call I was having fun, but I could tell something in my system wasn't right. I experienced bouts of severe nausea that were more serious than your typical repercussions of a wild night. I'd only gotten this kind of reflux once before, while at the Academy, and then it was because of extreme stress. I didn't recognize it at the time, but this was one of the first obvious symptoms that the strain of deployment affected me physically.

Pulling out of port, we only had a few more days of combat operations left before we turned over with the USS *Carl Vinson* and headed home. Each time one carrier finishes their work in the Gulf and turns the reins over to the next carrier strike group, the two mother ships pull alongside one another. All of the extra stores—bombs, missiles, and extra equipment for the jets—needed to be transferred. Also, the cross-deck allowed crews to turnover face to face, giving us a chance to pass our last words of wisdom and a few last-minute classified materials. From topside, I watched the operation—two one-hundred-thousand-ton ships steaming side by side, with helicopters flying back and forth transferring goods and passengers as fast as physics would allow.

Once we'd transferred the last load to the *Carl Vinson,* the *Bush* broke away and made a 180-degree turn to head south toward the Strait of Hormuz, our first obstacle on our long journey home. Once the passing of the torch was complete, the USS *Carl Vinson* would pick up the fight where we left off and the next day, they would be flying into Iraq and Syria.

Key Stats
Name: Caroline Johnson
Call sign: Dutch
Rank: Lieutenant (O3)
Aircraft: F/A-18F Super Hornet
Squadron: VFA-213 Fighting Blacklions
Total Flight Hours: 1,123.1
Brief & Debrief Hours: 3,607.5
Combat Flight Hours: 247.7
Enemy KIA (Killed in Action): 16 ISIS

Departing the Gulf and then the Middle East area of operations was called chopping out, and was marked by a change from tan flight suits, which matched the desert sand, to our standard US green ones. Naturally, we were all coming off a high, and already reflecting on the things we'd accomplished. Over the past eight months of deployment, our carrier air group had flown more than 32,000 flight hours, and of those, 18,333 were in combat. We launched 3,245 combat sorties into Afghanistan, Iraq, and Syria, and employed 120,000 pounds of ordnance. Our weapons accounted for 20 percent of the weapons employed thus far in the campaign against ISIS, and the weapons alone cost $62 million. We'd been the first aircraft to respond to the crisis in Iraq, had dropped the first bombs on ISIS, and led the first US air strikes ever into Syria.

The Blacklions, one of nine other squadrons aboard the *Bush*, had also done well. We'd led in the number of weapons we employed, we were the first to drop bombs in Iraq and Syria, and we set the standard for tactics and operations around the boat. As a squadron, each pilot and WSO employed in combat; more importantly we didn't "sling any bombs"—meaning, we hit all of our intended targets—and as a team, we didn't have any inadvertent border crossings or cause any international incidents.

These were all feats that other squadrons couldn't claim. Later, we would win many awards for our work, and even though there were three other F/A-18 squadrons on the *Bush* that year, all of the accolades for the Navy's premier fighter squadron on the East Coast went to the Blacklions.

When the *Bush* was relieved in the Gulf, our successors onboard the USS *Carl Vinson* would amp up the fight even more after we left. The release authority was finally pushed down from POTUS and senior leadership to a more appropriate level. During the *Vinson*'s eleven-month deployment, they flew 2,383 combat missions and dropped more than five hundred thousand pounds of ordnance. When they were relieved by the USS *Eisenhower* in 2015, things ramped up even more. Even though the *Eisenhower*'s air wing only flew nineteen hundred combat sorties, they dropped 1.3 million pounds of ordnance on ISIS.

These stats are more than just numbers, they tell a story. As the first show in town, the *Bush* kicked everything off and laid the groundwork for a long and grueling fight against ISIS. As time has progressed, the Navy got better intelligence and obtained more hostile targets on which they could employ. Even with fewer flights in combat, crews started dropping ordnance on almost every flight flown in-country. This showed that the US and our coalition were taking ground from ISIS and eroding their infrastructure. Even as an aviator who's been there and done that, it's still hard to believe we're still prosecuting that many targets, still dropping bombs. It's difficult to fathom that ISIS is still growing, like a weed popping up here and there, refusing to be destroyed, even after millions of dollars and pounds of ordnance and several years of work by the US and our coalition forces. No matter how many times the weed comes back, we can't stop taking action. We must continue to save the lives of those threatened by ISIS.

CHAPTER THIRTY-TWO

★

October–November 2014; Embarked USS *George H.W. Bush*, Mediterranean

Coming back from deployment we had the advantage of the long ride back, around the Arabian Peninsula, through the Suez Canal, the Med, and across the Atlantic. The trip seemed to provide most of us with the time to decompress, but for me, the return was actually busier and more stressful than the rest of deployment—finishing my master's, finding housing arrangements, and coordinating the logistics of life back at the beach. I was steadily managing things, but the strains of deployment began taking their toll before we even left the Gulf and had intensified by the time we got to Europe.

We stopped in the south of France for a port call on our way home and our squadron decided to spend our time in Provence. On this trip, for the first time since joining my squadron, Taylor did not want to share a room. She opted instead to stay at a château with some boys in the hills outside of Aix en Provence, a little town we'd planned to visit. Another group of guys and I chose to stay in the city, because it was quaint and convenient to the markets and local attractions, and while I found Taylor's decision a little strange, it was no big deal. We'd been together nonstop for nine months, so I figured a little space was probably good.

During our short visit, we hung out like usual, wine tasting in the beautiful vineyards, basking in the late fall sun, and enjoying the indulgent croissants and pungent cheeses of Provence. We enjoyed the cafes, shopped in the quaint, French provincial town, and on the last day, Carolyn, Taylor, and I met up for a girls' day. Getting back onboard, we hauled our cases of wine, French linens, and tired bodies back to the Sharktank for our last two weeks at sea. Overall, it was a perfect send-off for the Triple As, lots of delicacies, good times, and laughing, much like nine months ago when deployment began.

In addition to the souvenirs I'd brought back, I also carried with me an unwelcome memento from Europe—a nasty virus that mutated into a full-blown upper respiratory infection. When the flight doc said the virus would run its course, I took it as my invitation to keep grinding away. A few days later, when I landed from one of my flights, I got out of the jet and felt a shooting pain emanating up my leg and into my lower back, like I'd thrown my pelvis out of alignment. The force of all the catapult shots and arrested landings were catching up to me. Throughout deployment, I'd experienced back and neck pain that I'd managed with a foam roller and a strict workout regimen, but this was different, excruciating.

I was back at the flight doc's office, explaining how my whole back had seized up, and this time I left with some muscle relaxants and Motrin, but days later, with no improvement, the doc finally got me into the physical therapist for an adjustment.

Laying me out on his table, the therapist studied my crippled body for a minute. "Well, that's why. Your left leg is an inch shorter than your right," he said matter-of-factly, the way a scientist might point out the error in a long equation.

A couple of cracks and jerks later, my pelvis was realigned with my spine again.

"Better?"

"How did you do that?" I asked in wonder, experiencing almost instant relief.

"We see this in aviators all the time, particularly jets. I don't need to tell you, the flying—those heavy helmets and the shots and traps—are torture for the spine. You should be a little better now, but it's going to be a long journey getting your neck, back, and pelvis back where you want them."

Even though I was hobbling around like an old lady with a nasty cold, like everyone else, my thoughts were on our return to Norfolk. We'd departed on Valentine's Day, and now it was almost Thanksgiving. The ship hosted mandatory briefings to help everyone transition back into family life. I might not have been returning to a doting husband or darling children, but I was as excited as everyone else to get back. I knew my family would be there to welcome me, and there was a cute new apartment near the beach just waiting for me to move into it, and manicures any day of the week.

"If you experience bouts of anger or resentment for your loved ones," the reintegration counselor said, "get yourself to a safe place and wait it out. This, too, will pass."

Yeah, yeah, we've gotten through the hardest part, lady, I thought, totally unaware of the demons awaiting me just beyond the dimly lit pier.

CHAPTER THIRTY-THREE

★

November 14, 2014; Embarked USS *George H.W. Bush*, Atlantic Ocean

A day before the *Bush* pulled into port, all the squadrons flew our aircraft back to our home bases. The first squadrons to launch were those who had the farthest to go—Whidbey Island and Jacksonville crews left at the crack of dawn. The Blacklions launched midmorning, into the blustery rain. Our planned twelve-jet flyover was scrapped due to poor visibility, so we headed back two by two, a huge crowd anxiously awaiting our choreographed arrival back at the base.

My pilot and I had a more dramatic arrival than planned. Our jet, as if knowing its tour was up, landed and then caught fire, smoke and flames shooting from our oil-soaked brakes. Later, we figured out it was a common thing on Tailhook jets. On the carrier, the arresting cables brought us to a dead stop, so our brakes, unused for months, had accumulated grease and gunk that caught fire under the friction of stopping on a normal runway.

With the brake fires behind us and firetrucks headed back to their perch, we taxied to a stop in our line. Quite a large group of excited family and friends had come out onto the tarmac to welcome us home. My parents were there and so was Ashley, the WSO I lived with before leaving for deployment, and Virginia, my dear friend whose husband was deployed to Iraq. They held cute signs welcoming us home and cheered and ran to hug us as we climbed from our jets. I felt loved in the profoundly all-inclusive way one can only feel when coming home from a righteous war or an adventure of great magnitude.

As grateful and happy as I was to see my old friends, I had two people in my sights for my first hug. I didn't have to look far because as soon as I stepped off the ladder climbing down from the jet, there they were.

"You sure do know how to make an entrance," my dad said, slinging an arm around me. The lights from the firetrucks still spinning behind us, I embraced my parents.

"I can't believe we're finally home," I said, hugging my teary mom.

"Sweetie, we're so proud of you," she whispered.

Proud. I'd heard it before many times but this time it caught me. I was deeply grateful to hear it, even from my mother. "Thanks." I kissed her forehead. "Now let's get the hell out of here and head out into the real world. I have a new apartment and some online shopping waiting for me at the beach!"

The next day we all headed back to Norfolk to welcome the *Bush* home. It's an amazing sight, a one-hundred-thousand-ton supercarrier easing its way to dock like a prehistoric metal beast. On the flight deck, the ship's Sailors rimmed the edge of the boat while standing at attention in their service dress-blue uniforms in a tradition called "manning the rails." Thousands had traveled from around the country to welcome their loved ones home. Waiting on the pier for my Sailors and squadronmates to come ashore, I was surrounded by people who had tears streaming down their cheeks. Off to the side, I noticed a special enclosed area for the wives holding new infants, the first meeting for these babies with their dads. The whole atmosphere coursed with relief, and joy, with an undercurrent of uncertainty.

After the euphoric flurry of homecoming parties had ended, the routine of life and work returned. I'd moved into a great apartment only two blocks from the beach boardwalk, and I consumed myself with getting my life out of storage. After living with five others in a space that wasn't much larger than a janitor's closet, I'd decided to embrace my privacy and live by myself and worked hard to turn my apartment into a cozy beach flat.

Outside the home, in the romance department, I'd pretty much come to terms with the fact that Minotaur and I would never really be together in an adult sense. We'd been kids together and I cherished that time. No one had treated me better than Minotaur. Nor had anyone been as outlandishly romantic as that tough Marine. After all, he'd followed me to war based on the absurd and romantically distorted notion that he'd be someone between me and harm's way. Those kinds of thoughts are the romantic dreams of boys who dream in the language of a Chuck Norris movie poster. Lovable as he was, I needed more. We both knew it.

So I wasn't surprised when, despite his talk before leaving for deployment, once we were both back from deployment and the danger was over, I only received a random text or two. I was, however, a little surprised when I spotted a picture of him with his new girlfriend on social media, the two of them perched in front of a Christmas tree. Her caption read: *Starting some Christmas traditions.*

It hurt a little but not a lot. And once and for all, I said goodbye to the Minotaur and wished him and his girlfriend the best of luck with their Christmas traditions.

Beyond the Minotaur, I had harbored some hope that Burberry and I could rekindle a bit of the spark we'd left smoldering. Unlike Minotaur, Burberry and I had made no plans to get together, which now that I was back, made the idea even more appealing. I'd been debating texting him when, on a Saturday afternoon at a restaurant in Virginia Beach, I ran into one of the Jet Girls in his squadron. We ended up having lunch together, and although I had kept our relationship a secret, apparently he had not.

"So, Caroline," she said, "I've been meaning to tell you. Burberry. He was head over heels. He gushed about you nonstop."

This sounds promising, I thought. "Really, what did he say about me? He's a nice guy, but we're just friends."

"I don't know." She rolled her eyes. "He's a total player. He was smitten with you, though. He's getting married in two weeks."

"He is?" I said. "What?"

"Yeah," she said. "Clearly he didn't tell you about it."

"Yeah, I haven't talked to him since I got back. I didn't even know he had a girlfriend."

"They were broken up when he was at Topgun, but anyway. They're back together, engaged this time."

"Gotcha," I said, not necessarily surprised, but registering a blow nonetheless.

"Caroline," she said, watching my face slacken. "You're better off without him."

★

December 2014; Virginia Beach, VA

So my new life back in Virginia Beach, which had started out promising, began deteriorating rapidly. But a far more damaging loss than that of potential boyfriends was the end of my friendship with Taylor.

Taylor was also struggling with reentry and relationship

issues. We both were battling our own demons. Starting in France and continuing after our return, I felt Taylor cooling toward me. At the same time, I started to struggle again with some of the guys who bullied me before deployment. Snarky comments and tense interactions returned, as if the boat, the war, and the unity we forged in Iraq fighting together as Blacklions was something I had dreamed up.

When I moved into an apartment by myself, all I wanted was privacy, a clean space to not have to tape *Caroline* labels on the food I put in the refrigerator. I was after comfort, but ended up isolating myself. I can see the mistake now, but it was hard in the moment to understand that despite wanting space, I needed others around me. We all did. Some had families, the rest had each other.

Taylor, I think, understood this need and she surrounded herself with friends to help her through the return home. She moved into a beach house full of Blacklion JOs. In fact, it was called the JO house, and it operated as a sort of fraternity for the squadron. The vibe in the JO house definitely skewed "bro," with lots of drinking, joking, guys hanging, and pee on toilet seats. I didn't go there often, but when I did, I usually had a good time.

One night, I made plans to meet friends for dinner. A friend and I had finished our happy hour a little early, so to kill time before our meetup we decided to head over to the JO house for a drink and to see if Taylor wanted to join us later.

We arrived after most of the people in the house had gotten back from an afternoon at the beach. It was your typical JO pregame scene—dudes hanging in the kitchen, some shirtless, drinking beers; country music was playing on the stereo, people were in and out getting dressed, and I could hear ESPN on more than a few TVs in the house. Taylor was there and was acting a little cagey. As I hung around, trying to get a word with her, a couple of guys from the Academy showed up who made me uncomfortable. These guys were friends of a clique at the Academy who had gotten in trouble, and as a punishment, had gotten their graduation delayed.

I didn't care for the guys and I know the feeling was mutual. So I shrugged it off and headed out. During dinner, my phone lit up with a group text from one of the guys, the guy I liked least in our squadron: *Don't come by the JO house again unless you've been invited.*

The message was directed at me. It was a group text, but I was the only person on the chain who had been there and wasn't still at the house. If the guy who sent it had any balls, he would have just sent it to me directly. Instead, he shared the message with a group so he could hide in numbers, not confronting me one on one, but openly ridiculing me in a crowd.

Taylor, of course, knew I'd been in the house. But she had become close with the guy who sent the message, and at that point was better friends with him than with me. In fact, she might have been sitting at the kitchen island, drinking with the guy, when I had left the house. Worse, she should have known how much a message like that would hurt me, but she never reached out to let me know I was welcome, to tell me something like, "Caroline, fuck these guys; I live here, you're always welcome."

It's not easy to tell guys like that to stop. I know this from my experience at the Academy.

★

Spring 2008; US Naval Academy, Annapolis, MD

Winters in Annapolis are relentless—cold, gray, seeming to drag on forever. We call the stretch between Christmas and spring break the "dark ages." On campus, the trees are stripped bare and the wind blowing off the icy Chesapeake cuts through the warmest peacoat. The bundled-up black figures of midshipmen, scurrying to class with their heads down, are the only signs of life in the monochromatic landscape.

Around mid-March a little light seeps into the dark ages and by late April, there's no prettier place on earth. Dogwoods and cherry trees bloom, the soft air feels warm but not oppressively hot, and everything glows. The whites shine whiter, the greens brighter. And for the past twenty-five-some-odd years, a tradition, an annual rite of spring for the United States Naval Academy and the neighboring St. John's College, exists—a sprawling croquet match known as the Annapolis Cup.

On the croquet pitch, competitors from the traditionally liberal St. John's dress up in costumes, while the midshipmen from the ultraconservative Academy wear crisp white dress uniforms. Annapolis Cup spectators, locals, and those from surrounding areas, all come dressed up in classic 1920s wear—flapper skirts straight out of *The Great Gatsby* and hats suited for the Kentucky Derby. When I was a second-class midshipman (aka my junior year) I went to the cup and finally had the privilege of dressing up. I put on my best Lilly Pulitzer dress and Jackie O shades and watched the game from a picnic on the lawn. I drank champagne and ate finger sandwiches. Most of the guys were slamming beers and cocktails.

Since I had to drive that day to pick up an ice cream birthday cake for one of the plebes in my squad, I didn't let loose as much as I wanted to. I ended up keeping my drinking to around one glass an hour, and no more than three for the afternoon. I was definitely in the minority, even for Navy midshipmen. Many of us got completely sideways, and I was both happy for those who cut loose and hoping no one got in trouble.

As croquet wound down, I skipped out early to pick up the birthday cake. I arrived back at Bancroft Hall in time to change into my summer whites and meet my plebes in the hallway during chow calls before dinner. A chow call sounds something like,

"Sir, you now have fifteen minutes until evening meal formation, formation goes inside. The uniform for evening meal formation is summer whites. The menu for evening meal is jerk chicken, collard greens, mac and cheese, mixed salad, corn bread, blondie brownies, punch, sports drink, two-percent milk, and assorted condiments. The officers of the watch are: the officer of the watch is Lieutenant John Doe, first company officer; the midshipman officer of the watch is Midshipman Lieutenant John Smith, brigade assistant operations officer. The major event of the day is Annapolis Cup croquet. You now have fourteen minutes, sir!"

Outside my door were two plebes from my squad: Caitlin, the birthday girl, and her roommate. I stepped outside my room to surprise the girls with the cake when I saw two figures lumbering down the hallway toward us. One wobbled, the other helped him stay upright. As the pair approached, I smelled booze and recognized the wobbler as Greg, and the guy holding him up was Alex. Alex appeared relatively sober, propping Greg up and giggling at the drunken antics. These guys were not the finest midshipmen at the Academy. I had heard that, while the claims were eventually dropped, Alex had almost been kicked out when a girl alleged that he locked her in a bathroom and assaulted her in the bathtub at an off-campus party.

Caitlin, my plebe, was talking to me, oblivious to the guys behind her. "Birthdays at the Academy probably take some getting used to. Back home, there'd be a party and . . ."

"It'ssss your birrrthday?" Greg slurred, staggering up behind Caitlin into our personal space.

"Yes, sir." Caitlin turned, perky as ever. "My eighteenth!"

"Eighteen, huh . . . cool." Greg licked his lips and his eye seemed to twitch, a loose grin forming over his teeth. "I've got a present for you . . ." He slipped his hand down the front of his Nantucket red shorts. "Got one for you right here."

As Greg's hand fished around in his crotch, Caitlin let out

a weak "Ma'am?" and looked at me for help. Initially, I was too stunned to speak. Greg meanwhile intently struggled to pull out his penis, but he was either too drunk or too lacking in manhood to accomplish this task. Thus thwarted in his attempt to expose himself, Greg fondled himself under his shorts. I stepped forcefully between Caitlin and Greg, trying to shield my plebes from this appalling display.

"Jesus, Greg. Stop that!" I said, my voice authoritative but hushed, trying not to escalate the situation. Never mind, I'm sorry to admit, at that time I didn't want Greg to get in trouble. "Please, Greg, cut that nonsense out and go to bed."

Alex stood by, bizarrely unfazed, holding Greg by the arm. Without Alex's help, Greg would have fallen flat on his face.

"Dude, Alex, what are you waiting for?" I said. "Take Greg back to his room and put him to bed."

"Why?" Alex snickered. "He's all good."

"Good?" I said, awestruck. "He's being totally inappropriate . . ."

"No," Alex countered, as Greg leered at me. "He's not being inappropriate; what's inappropriate is you telling him to stop."

"Alex, you know what is inappropriate?" I asked, my self-control dissolving into total rage.

"What's that?"

"Cornering an eighteen-year-old girl in a bathroom and assaulting her." I leaned in a few inches from his face and stared into his flaccid eyes. "That's textbook inappropriate."

Even in his absolutely obliterated state, Greg seemed to pick up on what I was telling Alex. Greg's eyes bulged, shocked to hear me confront his friend like that, and then he fell into a drunken laughing fit. "Dude . . . she just . . . ha ha ha."

Alex turned as red as Greg's shorts. "You fucking cunt," he said. "You fuckin' drunk cunt. I'm going to have you breathalyzed. You fuckin' bitch!"

By now the altercation had drawn the attention of others from Alex and Greg's clique, mids who were cooler-headed. They rushed toward us, keenly aware of what would happen if Greg was discovered this drunk.

"I think that's enough for now," one of them said, and dragged the two back into Greg's room down the hall.

I turned back to Caitlin, who was still dumbstruck beside me. I managed a smile. "You girls go find another deck to finish chow calls."

They left, and I returned to a friend's room and flopped down. "Whelp, I'm getting kicked out."

"What happened out there?" she asked.

"No one talks about what they say Alex did, but they went too far . . ." I began to tell her what happened, but before I could finish the story, I heard a knock on the door.

Outside stood the breathalyzer crew. Blowing over a .08 could get me kicked out, and Alex, of course, knew this and he knew I'd been at the Annapolis Cup. He'd probably even seen me drinking champagne. He wanted to get me kicked out. He was out for blood.

"Okay," I said, stepping into the hallway. "Let's go."

I put the plastic straw in my mouth and blew.

The machine sounded and one of the crew looked at the readout. "It says zero," he said to the others. "Maybe it's a bad reading. We'll test her again."

"Sure," I said, keeping my eyes in the boat. "Best to be safe."

If the whole scene with Greg wasn't enough to sober me up, it had been four hours since the croquet match. I knew I was sober.

After I'd blown 0.000 twice, I'd had enough and turned my wrath around on them. "Now that you've breathalyzed me, you have to breathalyze Greg."

The breathalyzing crew now marched down into Greg's

room, where Greg was passed out so deeply he could not be awoken for several hours. Alex, brilliant strategist that he was, had not anticipated that I might do to his friend exactly what he tried to do to me. He started to panic, cursing me under his breath again. "Cunt. Bitch. Whore." But he couldn't stop what he had started. When they were finally able to wake Greg and administer the test, he failed miserably.

That night and over the next few weeks word got to me that Greg, Alex, and their posse were making threats that they would ruin my Navy career and find a way to get even.

The next morning I woke up to yet another knock on my door. Caitlin and her roommates were back. The same group of guys was physically blocking my girl plebes from using the women's bathroom.

"Follow me." I moved swiftly down the hallway and parted the group. I went straight to Greg and his friends, gritting my teeth. "Don't you ever try to intimidate me or these girls again."

Their gang backed off easier than I expected.

A formal investigation was launched into the conduct of Greg, Alex, and their friends. Initially the officer in charge took no action on the complaint, but eventually another witness came forward and pushed a new set of allegations. Finally the investigators learned about the whole story and revealed the extent of the harassment. An adjudication process followed. Alex faced no penalty, but Greg's conduct was found unbecoming of an officer, and his graduation was delayed.

While Greg and Alex were incensed by the punishment, in my opinion it was far too light, especially given the high standards of the Academy. The Academy exists not just to train leaders but to weed out those who do not belong in leadership roles. And I believe men like Alex and Greg have no business leading men and women in the United States military. Period.

★

Spring 2015; Virginia Beach, VA

Unfortunately for the Department of Defense, both Greg and Alex graduated from the Academy and pursued careers as Marine officers. Unfortunately for me, a few of Greg and Alex's close friends pursued naval aviation careers. In fact, one of them moved into the JO house after deployment. Whether intentionally or not, or as a result of the Alex and Greg incident or not, many of the senior JOs in that house undercut me and treated me poorly. As social influencers in our squadron, their behavior was mirrored by the other JOs in the Blacklions. And the worst part was that even my best friend, Taylor, eventually turned against me.

At Taylor's hail and bail, she gave a speech. I don't remember the exact words, but she talked about how fortunate she was to be part of an amazing group. She thanked almost every one of our squadronmates for their camaraderie over the years and during deployment. She even thanked several of the wives for their friendship and for helping her survive her tour with the Blacklions. She didn't mention me or so much as look at me.

I sat there in the crowd, waiting for anything, a smile, a nod, but it did not come. I remembered her words as she was soaring past me in the jet.

Everyone sucks but us. What happened to that friendship?

Taylor's words, or lack thereof, pushed me teetering over the edge of darkness into a bleak depression. Trying to pretend I was unphased and enjoying yet another never-ending mando fun event I spent the rest of the night numbing myself with the squadron's margarita machine.

And I wasn't the only one drinking 40-proof slushies. Goofy, soon set to marry, had basically funneled margaritas into his throat. His fiancée was out of town so he was cutting loose, and as we were leaving for the bars in Virginia Beach, one of the guys saw Goofy stumbling.

"Goofy can't come with us," he said. "He won't get into the bar. Somebody's gotta make sure he gets home."

With no other takers, I threw up a hand, annoyed at myself as I offered. For once, couldn't I just leave a man behind?

"I'll take him." I figured I would just have the Uber drop him off on the way home.

As we waited for the car, Goofy, true to form, drew a can of tobacco from his pocket and packed one in. He got in the back of the car and I got in the front, a precaution I took only because it was Goofy. The Uber had barely started rolling before Goofy was making his move. "Dutch . . . uhhh . . . hey, what do you really want to do tonight? Got any drinks at home? I bet you're a bourbon girl." He belched. "Excuse me. I love bourbon."

I tried to ignore him, chatting with the driver in hopes that Goofy might pass out.

He went silent for a few minutes and I thought my plan had worked, then without warning, Goofy reached up over my shoulders and started groping me from the backseat.

"What the hell?" I screamed, knocking his hands off my chest. I could smell the tobacco in his lip.

"Come on, Dutch. You know you want it. Just come home with me." He tapped the driver on the shoulder. "Bro, we'll just be making one stop."

The driver looked over at me. My icy stare said enough.

"Okay," the driver said, "that'll be two stops."

Goofy was a slob and a douche and he'd grabbed my breasts, but even with his despicable behavior, still, I dutifully waited until my squadronmate fumbled with his keys and fell inside his apartment before I let the driver pull away.

★

The final blemish to my return home was, in reality, a blemish. For the first time since I was fourteen, I had acne on my cheeks, nose, and chin. The dermatologist blamed it on the oxygen mask irritating my skin, but I was also having headaches, dramatic mood swings, and even though I'd been back to healthy eating after making do onboard the ship, I was still bloated and gaining weight.

By now, Carolyn had moved into the spare bedroom in my apartment for a few months while she was getting ready to transition to her next duty station, which in her case was the Naval Air Warfare Development Center. Carolyn, the daughter of a pilot who flew fighters, who had fought relentlessly to prove herself as a naval aviator, who had been given the call sign TATL by the guys in her squadron as a swipe at her very existence, was going to the most elite training cadre in aviation. I couldn't be more proud of her. Or scared to lose her. She kept me in check and helped me through the day to day.

She was experiencing almost the identical symptoms, and before too long, we did some research and found out that while we were on deployment, the Navy had substituted our previous birth control pills with a less expensive, identical-*looking* generic. After Googling the new brand, I discovered an entire message board where hundreds of women were experiencing acne, weight gain, and mood swings. I brought the issue up with our flight doctor, who interrupted me midsentence.

"Dutch . . . you're being a bit of a brand snob here. Sometimes people *think* something is better because it is more well known . . . or more expensive." He continued his condescension while I quietly fumed.

This small instance was just one thing in a litany of misunderstandings I was facing at the time. I was struggling, and as I normally do when I'm faced with adversity, my analytical, logical brain looks at the problem, sees what needs to be fixed, and then I take action. But in the moment in my squadron, the formula just didn't fit. It was as if I were back on the flight deck, standing there in harm's way, and no one would tell me which of the foul lines were activated. I kept trying to find a safe place just to exist, but everything was coming straight at me. I looked inward to see if the problem lay within me. I justified some of the treatment. I grinned and bore it. I plastered a fake smile on my face every morning and went to work, preparing for another deployment. Mostly, I could hold it all inside, except for the few times in the day when I would escape to the bathroom and give myself a few minutes to cry in a locked stall. Eventually, things got so out of control that my emotions would boil over at the most inopportune times. Pulling on my helmet, I hurried to the jet before anyone would notice my tear-streaked cheeks under my visor. Before I left for deployment, I couldn't will myself to cry when saying goodbye to my mother. And now I couldn't hold it together, as I was climbing into my jet, no less. What was happening to me?

In addition to the ongoing struggle with back pain, I had been evaluated for significant hearing loss from the deafening jet-engine noise. Physically, emotionally, and mentally I was falling apart, and I didn't even realize how depressed I had become. This was a kind of dark, deep-down depressed that I'd never before experienced.

Then Carolyn left for Nevada and I dropped deeper. I needed a friend. Despite feeling abandoned by Taylor, I still sought her out. I wanted her friendship back, but she scrupulously avoided me. Our only conversations were professional. I even struggled to make eye contact with her. At one point, I went to a close mutual friend to talk to her about Taylor. "What did I do to alienate her?" I asked. She had no answer; not one she shared, anyway. I pressed her. "Please help me to fix this."

"I don't know. I don't understand what's going on, that's between you and Taylor," she said, looking away uncomfortably. My other Jet Girls, too, started making less and less time for me. And while they never abandoned me as a friend in the same way Taylor had, I began to feel like my friendship had become a burden to them. When close friends have a falling-out it's like a divorce, the couple's shared friends have to choose which friend to stay loyal to. So I peeled away, isolating myself from most of them.

★

After flying all day, I would go home every night, and almost as soon as I drew my first breath of the stale air in my apartment, a sense of profound loneliness would envelop me like a lead blanket. I yearned for someone to be there to help pull the blanket off of my shoulders, to hold me, even someone to simply sit with me, next to me, on my empty couch. My family and closest friends were thousands of miles away, but it wasn't like I wanted a boyfriend because, in truth, a romantic relationship at that time would likely not have been possible. I've never been the type who needs a man just to fill a void, but at that time I just needed support and help.

Instead I was left alone with my thoughts turning destructive. I tried to work myself out of this deepening pit, focusing on my master's coursework. But you can only stare at a computer or pore over the pages of a textbook for so long. Eventually you have to look up and acknowledge your world is upended. And that's when the weight of the depression would fall back on me.

I'd curl up alone in my bed, flip on the TV, and stare at the screen. Anything to distract me from my thoughts. But there was no escaping them. My mind would race and I would lie on my bed fantasizing about ejecting at upward of six hundred miles an hour, or stepping in front of a speeding car on my walk to the beach. Unsure of what to do, worried that I might act on these impulses and wanting the thoughts to stop, I would crank the volume to its highest, trying to drown out my thoughts in white noise that sounded like jet wash.

If I'd been in a corporate job, or anywhere else other than the military, my letter of resignation would have been sitting on my boss's desk in the morning, but in naval aviation, you don't quit. Not only that, but as an aviator, you never admit you are depressed. That's not to say I didn't make a few cries for help. I had already tried to explain to the flight doc that things weren't right.

"Man up, Dutch," he'd finally told me, annoyed that I'd done my own research, trying desperately to be my own advocate. "Sometimes you've got to stop being a drama queen and just deal with it."

I wasn't just stressing about the present; I had future plans to worry about as well, namely, where I should go for my next duty station. As a WSO, I was committed to serve for six years after my winging, so the Navy still owned another two and a half years of my life. Like they would for any top performer, senior leadership expected that I would head to Fallon for training at Topgun or train new WSOs at the RAG or in Pensacola. Finishing the last of my master's in leadership, I'd discovered that I was experiencing textbook examples of failures in leadership. After reading case studies and research about hostile work environments, I actually wrote my thesis centered on my experiences. Realizing I was at my wits' end and I wasn't going to survive my last year in the squadron, I finally decided to reach out for help.

Time had rolled around for the JO annual performance reports, so I already had a meeting set with Chick, my commanding officer, and Beans (short for bean counter), our XO who'd been with us eight months. As expected, the review went quickly, and they told me I was doing a good job, but that's where the positive reinforcement stopped.

"Dutch," Beans began. "We've noticed that you don't quite seem like yourself lately."

My fake smile faded and I assumed my blank stare, nodding in agreement with their assessments.

"It seems," he went on, "you've been isolating yourself. And we just think you'd benefit from trying harder to fit in with the guys."

At that point, I couldn't take it anymore. I had to tell them what was going on.

"Well, gentlemen, the problem is that I—" I broke down sobbing, explaining how I'd gotten to this point of isolation and how truly miserable I was.

They couldn't believe what I was telling them, until I gave a few specifics.

"Maybe you could talk to the squadron?" I said to Chick. "You know, ask them to bring me into the fold, kinda like you did for Cupcake?" It was humbling, of course, but I just wanted in.

"Dutch," Chick said after a pause. "We really had no idea this was going on. I'm so sorry."

In that moment, I was drowning in a vast gulf, but seeing the compassion in his eyes felt like someone had thrown me an orange life ring.

"Let me ask you . . . what are you thinking about doing for your next tour?"

"My shore tour?" I asked. *Why was he bringing up my shore tour now?* It took me a second, but then I realized what must be happening. Before they would stand up for me, I figured, they wanted an affirmation of my loyalty to the fighter community, they wanted to know if I would stay on the golden career path and promote in the Navy before they would stick their necks out.

"Well," I said, trying to regain composure. I knew I had to be careful. In the Navy, and especially in jets, if someone catches a whiff that you're not all in, your career is doomed.

"Well, what's it going to be, Dutch?" Beans pressed.

I cleared my throat. "I'm all in, sir."

CHAPTER THIRTY-FOUR

★

Fall 2015; Virginia Beach, VA; Las Vegas, NV

In my heart, I didn't know how much time was left in me. As the months ticked on, I slowly started to realize that I wasn't where I needed to be. Even though I loved the flying and all the challenges, I'd long ago stopped having fun. I love to work hard and play harder, but I'd finally come to the realization that as a woman in the fighter community, I was expected to give at least 200 percent, while knowing I'd probably never be ranked or fully accepted as an equal.

Maybe it was that my personality wasn't a fit, maybe I committed too many faux pas as an FNG in the squadron, or maybe some of the bros were right and I was just an ice-cold bitch. Maybe, maybe, maybe. But through my master's degree and through living, I'd realized a few things.

First, I have a big personality. Assertive, strong-willed, gregarious, and willing to take risks, I was raised to be open to new experiences and have self-confidence, all prerequisites necessary to survive in a fighter squadron. These qualities, while highly valued for males in my field, when exhibited by a woman, were despised by some. Often women with these qualities are labeled as bitchy, aggressive, or frigid because these traits are not typically thought of as feminine. As an officer and aviator in the fighter community, in a very nontraditional gender role, I needed those characteristics to survive, even though they subconsciously aggravated many of the men I worked with.

By the same token, the aviation community is openly family-centric, in the most traditional, 1950s way, and I was realizing that I'd never be accepted in this family setting as long as I was single. Which was also a double-edged sword. It's hard enough to meet a guy with the kind of values and qualities I was attracted to, but the squadron lifestyle and constant coming and going made it difficult to meet anyone outside the military. Occasionally, though, I'd catch a glimpse of what life could look like on the outside, even if a fleeting glimpse.

I'd been training for two and a half weeks near Las Vegas, working with Air Force and Army units. During the day, we'd fly and at night we'd hang out on the strip. One night at the end of the trip, my friend Carly, the WSO who'd replaced Taylor in the squadron, and I decided to branch off from the guys (who were headed to a "titter") and go to a Britney Spears concert. Carolyn, who since our deployment had gone to Fallon, came down to join us. She didn't come for the music or the fun time, but for me. She knew I was slipping over the edge of an abyss and she did everything she could to pull me back. She was the only person in the Navy I told that I was considering leaving jets. I told her when we were alone in our hotel room, getting ready for a night out on the town. To my shock, she took me aside in Vegas and told me, "Caroline, you can choose a different path. You *should* choose a different path. You can't let this bullshit go on. Fuck them all." I struggled to process the conflicting emotions coursing through me. Carolyn had reached the almost impossible heights that I, and every aviator in F/A-18s, had at one time aspired to. She'd made it, she was on the golden career path. And here was this amazing woman I so admired, the friend who has stood by me when others left my side, telling me it was not just okay to quit, but a good thing. I could only nod and mouth the words *thank you*. She turned away and looked at me in the mirror, smiling now. "So now we forget about that. Let's go have a blast, and don't say shit about what you are thinking to anyone else or they'll ruin you."

Entering the auditorium the ushers identified our military discount tickets and we were pulled out of the crowd and escorted to upgraded second-row seats. Cheesy as it sounds, the concert was awesome. We sang our hearts out to the Britney hits we'd grown up with.

Afterward we went to The Chandelier, a luxurious cocktail lounge in the Cosmopolitan hotel. And for a brief spell I was able to turn back time. I felt like we were on deployment doing a port call before things went really bad. It was Caroline and Carolyn together again. At some point, an almost absurdly handsome forty-something guy came toward the table next to us. He smiled a little sheepishly, like he was heading our way on a dare, like a husband away for the weekend with his wedding ring in his toiletry bag. I was apt to ignore him when he spoke with a British accent. "Hi, who *are* you, girl, and what do you do?"

"Hi, I'm Carolyn, I fly F/A-18s."

Nod to me. "I'm Caroline, I also fly F/A-18s."

"Really," he said, then squinted, making a puzzled, dubious expression. "This isn't a joke, is it?"

My look said exactly what I was thinking and I was set to

turn my back on him when he added, "Because if it's a joke, I'll be seriously disappointed. I've never met anyone who's flown an F/A-18. And right now you've changed everything I've ever thought about the military."

"Oh," I said, "what did you think?"

"Let me buy you girls a drink and I'll tell you."

It turns out this gorgeous Brit was a marketing executive, in Vegas not for a bachelor party but for work with one of his major clients. After drinking and talking late into the early-morning hours, he invited me to dinner the following day. He had to take a red-eye out that night, so he made it clear that all he wanted was company and conversation.

"Sounds good," I told him.

As he had business in Vegas, he had been there often and knew the city's secrets. He started off the night with a tour, showing me some of the gems he'd discovered, like the Paiza Club on the fiftieth floor of the Palazzo, a private gambling establishment that caters almost exclusively to Chinese high rollers. It felt like Beijing, the TVs turned to Chinese stations, the newspapers all in Mandarin, and private gambling rooms with tables that had stacks of $100,000 chips. "The Palazzo will fly you over from China in a private jet if you guarantee you'll gamble a minimum of one million dollars a week," he told me.

"Sign me up," I said.

We ended our night with an intimate dinner at STK, an ultramodern, dark, yet inspired restaurant. When the check came, he snatched it off the table before I could offer to pay. "It's on me tonight."

I caught a quick glance at the bill. *Six hundred dollars.* My head spun. *Eight with tip.*

"Caroline," he said as we left STK. "You know I'm leaving for London tonight, but I have my room for the rest of the week.

It's yours, if you want it. Just let the butler know how long you'll be there." He put the key into my hand and tried hard not to smile—the Cosmopolitan.

His room was a twelve-hundred-square-foot suite with wrap-around balcony on the thirty-fifth floor.

"Yes!" I threw myself on the freshly made bed, purring with delight like Julia Roberts in *Pretty Woman*. Showered and under the covers, I looked out at the city, spread before me, twinkling in the night. I laughed out loud, thinking somewhere down below, the guys in my squadron were in a crummy titter hut, jamming dollar bills into G-strings.

Landing from our trip, my pilot and I parked next to Chick's jet where his wife and boys were waiting to welcome him. It was one of Chick's last big trips in the squadron, so his homecoming was celebrated with a bit of fanfare and a postcard-worthy family reunion. The handsome aviator strolling toward his dutiful wife holding back tears, her children behind her, warm pink light from the setting sun cast across their expectant faces.

As we all grabbed our bags, the skipper's wife peeled away from her picture-perfect family and headed over to me. My flight bag, a clunky maintenance book, and my overnight bag were a lot on my back, so I was focused on getting to the squadron as fast as possible when I heard, "Looks like you had fun in Vegas."

I turned to Chick's wife. "Yeah, we did, thanks!" I said and kept walking.

"Well, I saw pictures of you on Facebook, and you looked really tired."

Where's she going with this? I thought, then told myself, *fuck it. I can't play a game right now. Gotta be direct.* "We looked tired?" I stopped and squinted at her, her face now appearing red and agitated in the pink sunlight, like there was lava under her skin, boiling. "What do you mean by that?"

"When you girls stayed on the strip, did you sleep with the guys?"

"Excuse me?" I readjusted the bulky bags on my shoulder and checked her expression to see if she was serious. "Oh, you mean, did we sleep in the admin suite with the guys? Or did you mean something else?"

Her eyes flitted over to Chick, then back at me. "No, I mean, did you share a suite?"

I was tempted to lie and tell her yes, to confirm her suspicions, if only just to see what would happen to her facade when whatever was boiling underneath the surface erupted in front of Chick and in front of her kids. A small part of me wanted to see her crumble right there in the middle of her own fantasy reunion.

It might have felt good to do that—to push her buttons, the ones she was begging me to push. But that's not me. So I told her, "Absolutely not. With all due respect, I'm not a college kid. I have no interest in sleeping in an overcrowded suite with twenty or thirty of my colleagues." I wanted to add, *or go to strip clubs with your husband and the rest of the dudes every night for the past two weeks.* But I just said, "Carly and I got our own room."

"Okay," she said, pursing her lips and tousling her hair. "Welcome back."

★

After the chat I had with Chick and Beans, I started to notice that even though I was still telling everyone I wanted to stay in the fighter community, I was quietly being sidelined. When Beans orchestrated the semiannual JO job change, instead of keeping me on the golden path, assigning me a more demanding job, he gave me a ground job that was typically forced upon the newest new guy in the squadron—public affairs officer.

And even though I'd mastered my last qualifications as a combat division lead and mission commander, I started noticing I wasn't scheduled to fly any prime events. I was one of the few JOs qualified and capable of teaching and leading, but I was consistently put on odd-houred flights and events that weren't tactically enhancing.

As these career slights occurred more and more, it was clear—I was being iced out. But at the time, I was so beaten down, I didn't care. As we say in jets, gas is life, and as I studied my own emotional fuel gauge, I knew I was running on fumes. If the Blacklions seemed not to want me, I would just steer clear. I would set max endurance, conserving what little energy I had left, spending as little time at work or interacting with my squadronmates as possible. My peers reaped the benefits and absorbed all of the good deals dealt out by Beans. I didn't blame them. In the end, we were all competing against one another for our rankings, which would determine our future jobs and affect our futures in the Navy. The guys needed the harder jobs, better flights, and more prestigious collateral duties so that they could be ranked higher than me.

Around Beans's one-year mark in the squadron, I was as good as invisible to the Blacklions. My work didn't matter, I didn't matter, and no one would have known or cared if I was there one day or gone the next.

The steady *drip, drip, drip* of criticism, insults, and slights had eroded my confidence. The effect was not just superficial. Like a constant trickle of water on limestone over time can wear away bedrock to create an underground cave, my confidence washed slowly away, and a pit formed inside me. Where I once felt strong and steady and was literally ready at a moment's notice to rush into war at Mach speed, I now felt empty, fragile, and defenseless.

Even though it was the first time in my life I felt like I had no fight left in me, there were two things I wanted to accomplish. The first was finishing my master's degree. Channeling all of my work frustrations and energy into finishing my coursework and using many of my experiences in the squadron, I wrote about leadership in my thesis and completed the master's with a 3.72 grade point average.

The other thing I wanted to do was join a Navy business trip to Denmark. Our squadron had been tasked by higher ups in DC to support a Boeing trip to an air show in Copenhagen. We would be there to help sell the F/A-18 to the Danish and other potential European buyers.

Denmark had contracted to buy the F-35 joint strike fighter years prior, but because the program was so plagued with delays and problems, the Danish government was looking at other options. Our Super Hornet was one of the top contenders. When our squadron found out about the opportunity, Beans put out an email requesting volunteers for the trip. I responded immediately with a note expressing all the reasons why I should go. I was the senior-most JO at the time, I had experience from the Academy in foreign affairs, and I was fluent in German. I knew there would be German press there, and as the squadron public affairs officer—the crappy job Beans had pawned off on me—I was the person designated to represent our squadron at these types of events. A few days later, when I heard rumblings among the officers about who'd been selected to go, my name wasn't in the mix. I mustered every ounce of fight I had left in me, went straight to his office, and knocked.

"Sir, you got a minute?" I asked and he motioned to a chair. "Sir, I was just wondering . . . I've heard rumors that the Boeing trip has been decided, and my name isn't on that list."

Beans didn't look up from his paperwork, so rather than waste time with small talk, I dove right in. "Of all the JOs, we both know I'm undeniably the most qualified. So I'm just wondering why . . . I'm being passed over."

"Dutch," he sighed, pushing back in his chair. "I can appreciate your desire to go, but I don't think this is the best job for you right now. I've already assigned other crews that will be going."

"But, sir, this type of trip is exactly where I can do my best work. I'm the perfect person to talk about the capabilities of the Super Hornet. I've employed her in combat, dropped bombs, and employed live missiles. I'm articulate, fluent in German, and I've dealt with high-level European diplomats and leaders on deployment."

"I really do understand what you are saying, Dutch, but I've already made up my mind."

The conversation went back and forth. I had one ace up my sleeve that I didn't want to use. I wanted to earn this trip on merit. Because I was the right person to join the team, but when all of my arguments and reason fell on deaf ears, I pulled out the trump card.

"Sir, in addition to everything I've told you, I also deserve to go on the trip."

"*Deserve?*" Beans said. I had his attention now for sure.

"Yes, sir. When we were on deployment for nine months, I was the only aviator in the squadron who was not sent on a detachment." Detachments are like short breaks from deployment; usually aviators are allowed off the ship to join one of our allies at their home base to fly and train for three to four weeks.

"Well, Dutch, that's because you wouldn't have liked living in a tent in Oman or Djibouti or any of those other countries. Also, women weren't allowed on most of those trips."

"With all due respect, I would've been just fine. No one ever asked if I wanted to go. And actually, there *were* females on those detachments." I found it funny how quickly Beans forgot that I lived with Carolyn and women from other squadrons on the boat who'd been chosen for the detachments.

"I deserved to go, I asked to go, and yet you and skipper never let me," I said, feeling like a hotshot litigator delivering closing arguments. "I haven't said anything until now, and to be honest, I didn't ever want to bring it up, but it isn't right. All I ever expected or wanted in the Blacklions was fair and equal treatment, and so far that hasn't happened."

Trying to hide his sheepish look, he cut his eyes over to his computer screen. "Look, Dutch. I can't change how you feel about the squadron. Sometimes that's just how life is. You can't always get what you want. The list is decided."

Furious but contained, I said coldly, "Roger that, sir. Just want you to know that I deserve this opportunity before I leave the squadron. And if I don't go on this trip, this won't be the last you'll hear of it." I did not need to remind Beans that the top brass had a mandate to ensure the Navy at least appeared fair and equal among the sexes. I had learned in combat that when the enemy threatens you, you must be ready to escalate to the next level of force.

At that point, Beans must have known what would happen if he didn't fix his mistake. Clearly, I felt, he wanted to put me out to pasture, but he must have known he had to cover his own ass. I'd never played the gender card before, but I was so emboldened at Beans's audacity to deny me an opportunity, I laid all my cards on the table. I don't know what went through his mind after I brought up the detachments and made my threat, but I can tell you what he knew about me. One, he knew I could do the work. I knew the business as well as anybody else. Period. Two, he knew that I was aware of his penchant for playing favorites and by confronting him, I wasn't afraid to stand up for myself. And three, deep down I think he knew I was right, and it wasn't fair. I deserved to go.

As I turned to leave his office, he took a long breath. "Okay, Dutch, let me see what I can do."

CHAPTER THIRTY-FIVE

★

August–September 2015; Bangor, ME; Copenhagen, Denmark

Six hours after launching out of the northernmost military air-field in the US and soaring high above the icy, white-capped waters of the Arctic, we entered Danish airspace. Snuggled up tightly on our lead jet, we peeled off the massive Omega tanker that had safely escorted and refueled us for the duration of our long transatlantic flight. It was late afternoon and the shadows were growing longer as we descended over the beautiful Danish farmlands that looked like a hand-sewn quilt of rich, Indian summer colors. When we checked in with the tower at Roskilde—our final destination—the air-show director requested a low-fast flyby of the field before we came in to land. We knew there would be some Boeing support crews and air-show organizers there to meet us, but we had no idea the degree to which the press and public would also be involved.

As we landed and taxied up to the ramp, a large crowd had assembled with a sizable press corps. Reporters love pilots, so WSOs usually take a backseat, both in the plane and when talking to press. So when I popped open the canopy and stood on the wing to stretch my legs, I was glad to know that of the four of us, I would not be in the spotlight. But as my helmet came off, my braid came loose, and my blond hair spilled out across my shoulders, I could hear a murmur ripple through the crowd. I didn't know it at the time, but there was only one female jet pilot in the entire Danish air force, and if you've ever been to Scandinavia, you know that Danes and Scandinavians in general are some of the most progressive people on earth. I'm sure they expected to see a square-jawed American military man with a buzz cut, and yet the first person getting out of the arriving jet was a tall blond woman that looked like them.

I climbed down the ladder, and the next thing I knew, three cameras and an ice-cold can of Carlsberg were thrust in my face.

"Open it." The reporter nodded at the beer with a smile. Parched from dehydrating myself so I wouldn't have to pee during the flight, and not wanting to be rude, I obliged. I popped the top and took a nice long swig, not even thinking about the fact that I was on camera. I answered a few of their questions, telling them how excited we were to visit their country, and I posed with my pilot and our plane for a few more shots.

"That was quite the welcome! No one warned us to expect such a crowd on landing," I said as we piled into the Boeing van headed to our hotel.

"Oh yeah, the Europeans go crazy every time we bring jets over." Boeing's media director chuckled. "Your arrival and interviews are going to be on the ten o'clock news tonight. Did you see the helicopter filming you when you buzzed the tower?"

"What? They were airborne for our flyby?" I was surprised that I didn't see the helo, but then it dawned on me that they were flying above us because we were so low, and I was only looking for contacts on or below our altitude. "If I would have known about the welcome party, I would have at least touched up my makeup before climbing out of the jet. The oxygen mask is killer on the complexion," I joked.

"No need to worry," the Boeing rep said. "There will be plenty of opportunities for you to shine, Miss Duchess," he said, misconstruing my call sign "Dutch" and giving me a knowing wink. I could tell we were going to have a great trip.

But it wasn't all play. We showed off our jets at the air show and spoke with the Danish fans about the F/A-18's capabilities. We worked the different events and stood safety watches in the tower for our demonstration and VIP flights. To my surprise, I was asked to help Boeing's executive vice president of military aircraft sales and Boeing's lead test pilot talk to the Danish defense minister about our beloved jet.

We spent almost an hour escorting him through the simulator and showing him the F/A-18 Super Hornet. The Boeing VP did most of the talking, but the Danish minister had some doubts about the Super Hornet's performance in combat and was concerned how well the jet performed in combat and against high-level air-to-air threats like the Russian fighter jets.

Before the VP could answer the minister, I interjected. "Ma'am?" I asked while stepping forward in the circle that we were standing in. "If you don't mind, I think I can speak to the minister's question."

"Dutch, absolutely," she said, turning the stage over to me.

I thanked the Boeing VP and turned my attention to the minister and his entourage. "Sir, I recently flew this jet in combat in Afghanistan, Iraq, and Syria, employing air-to-ground ordnance—specifically bombs and missiles—so I have experienced its capabilities and versatility firsthand. The two-seat setup with the current software package and sensor configuration make the Super Hornet an incredibly effective weapons platform in opposing the current threats we're facing in the Middle East. The systems are intuitive and combat-proven and enable crews to seamlessly integrate with all the coalition forces to quickly and accurately deliver weapons on dynamic targets.

When prosecuting ISIS targets, we were as fast and lethal as any other platforms out there, regardless of rank or experience of their aviators versus ours."

"How quickly can one of your aviators complete training and then employ weapons in combat?" the minister asked.

"The day our pilots and WSOs complete our F/A-18 training syllabus, they are worldwide deployable and are capable of flying in combat and unleashing devastating effects on the enemy. I think the Danish would benefit from integrating into our existing F/A-18 training programs like the British are."

"Interesting . . ." He thought for a second. "What about air-to-air? We are having major issues with aggressive Russian fighters, and I've heard that the F-35 is the only jet that can go against them."

"I also have extensive experience with air-to-air engagements as a combat division lead and mission commander, so I can understand your concerns about the Russians. Without getting into classified aspects of the jet, I can tell you that we have an upgraded radar suite in the Super Hornet that is extremely capable. We train to fight the highest-level threats and are consistently able to detect, lock, and shoot the enemy before they even know we're there. Daily, we smoke our most capable adversaries in realistic training scenarios. Trust me when I say the Russians are no match for the F/A-18's technology or tactics."

After I answered a few more questions, we wrapped the meeting. Afterward, back at the hotel, I realized I had commandeered the latter half of the discussion and was worried that I'd said too much. I tried to shrug it off and went down to the bar to join the Boeing crew.

Should have shut up, I told myself, and ordered a martini, angry that I had let my guard down and opened up. I thought about Beans somehow hearing about this lapse, and envisioned him chastising me about overstepping.

I was silently berating myself, hunched over the bar, starting my cocktail when I heard my name.

"Dutch." I turned to see the VP from Boeing approaching me. The woman—yes, woman—who ran sales for all of Boeing's military aircraft, including the F/A-18 Super Hornet. "That was incredible what you did earlier," she said. "You were spot-on when you spoke to the defense minister."

I had grown so accustomed to doubting myself that it took me a moment to process her words and recognize the compliment. "Thank you, ma'am." I sighed, relieved. I sipped my martini and instantly felt the muscles in my back loosen. "Glad I could help."

"But I want to know, how did you know what to say?" She put one hand on her hip and gestured with the other.

"I'm not sure what you mean."

"No one from our team prepped you or told you to say those things?" She stepped toward me and lowered her voice. "I'm in sales. I do these meet and greets all the time. No Navy guy we've ever pulled up has performed as well as you did. You said exactly what the minister wanted to hear. You broke it down so he could understand the jet, and you relayed the capabilities better than most people who've been in the business for years."

"Well, thank you, ma'am." I straightened up at the bar. "I'm passionate about these machines. The Super Hornet is an incredibly capable jet. I love flying it, so I just told him the truth."

The executive took a little breath in and started to say something, and then stopped. "Thank you . . . thank you for being honest."

A month later I heard from the VP, telling me that our conversation with the defense minister turned the tide of the deal, and Denmark was reconsidering buying the F/A-18. And that if I ever wanted a job with Boeing, she would have me.

CHAPTER THIRTY-SIX

★

Fall 2015; NAS Oceana, Virginia Beach, VA

We draw moral and ethical guidelines for ourselves in an attempt to set standards that we believe we should meet, standards that define us and guide our conduct. For me, long ago I laid down two such lines in thick permanent marker: I would not engage in a sexual relationship with a man in my squadron, nor would I engage in a sexual relationship with a man in a committed relationship, be it marriage or a girlfriend back home, wringing her hands hoping the guy she loves comes back alive. Those lines could not have been clearer. And no matter how high the stress, sexual tension, strength of the cocktail, or charming the behavior got, nothing could blur those lines. They mattered to me because I felt not crossing those lines preserved some special dignity for me that no wife nor whisper behind my back could take away. I felt that by not crossing those lines I could always stay above criticism, innuendo, and avoid looking like a girl who slept her way into the backseat of an F-18.

Those lines, however, were drawn when I was strong, before I had a gaping emptiness inside of my chest, back when my defenses were still up. It began innocently enough.

I was in a sushi bar in Virginia Beach, staring at a plate of uneaten nigiri, drinking my third glass of Sapporo as slowly as I could. I didn't want to eat and didn't want to get drunk (though the idea was getting more appealing with every sip), but more than anything I just didn't want to return home to leaden loneliness waiting for me inside my apartment. I knew from the JO group text message that there was a party at the JO house, but given my treatment the last time I was there, I would not be dropping by uninvited, or invited. I had picked the restaurant not for the food I wasn't eating, but because I was hoping I might run into some Navy SEAL friends. They often hung out there when not on a mission. I'd seen them a couple of times before and we'd talked and exchanged numbers, but I didn't want to call or text. I wanted to bump into them.

Virginia Beach is like that. One minute you think you're in a sleepy seaside town packed with retirees, and then you look up from your bowl of edamame and there's a handsome guy who's jacked like an NFL running back sitting next to you at the bar in a pair of shorts and flip-flops, and you know that less than forty-eight hours before he could have been on the other side of the planet snatching a terrorist out of a heavily guarded compound.

I found myself drawn to these types—not necessarily SEALs per se, but those who, to one degree or another, lived on the far edge of the spectrum and had shared some of the intense experiences I had, or at least had some context. I have never dwelled on the lives that I had taken, but I'm neither proud nor regretful. I did it to protect the innocent lives of those who couldn't protect themselves, and because it was my job. That said, part of my experience included killing sixteen terrorists. That's not a normal thing to do outside or even inside of the military. Nor is the way I killed the men normal. I used technology that allowed me to essentially reach down with my hand and swat the ISIS soldiers with a fistful of fire while tearing through the sky at five hundred miles per hour. Guys like the SEALs and the Jet Girls who've fought in combat understand that even the most well-adjusted of us are affected by what we do. There's baggage that comes with it. It's neither good nor bad, but it's part of me. Some people know how to handle it, others don't. Sitting in the sushi bar, I wanted to be around those who get it.

Perhaps more impactful—at least in my case—was what I witnessed ISIS do to other human beings and realized the profound influence my experiences had on my worldview. After watching ISIS take human lives for granted like they did and after more fully understanding their ideologies, I have become less sympathetic to those who intellectualize the US's role in the Middle East. I feel justified in what we do, have done, and what we need to do to finish, to keep America safe from the dangers we cannot even fathom. Knowing what life is like for both the citizens in these war-torn countries and for the men and women who are fighting to help bring order, makes it hard to come home. I struggle to connect with people whose biggest problem is a long line at the grocery store. And I have little patience for those who endlessly complain about our leaders, policies, and our country without having done a single thing for it, save grudgingly paying taxes every year. Once I saw the magnitude of the true evil that exists in the world, it made me more accepting of taking a hard line with our enemies and enacting careful screening mechanisms to help ensure those with evil intent and the capability to inflict mass harm stay off of our soil. Having this sort of worldview in most circles means I either need to debate, hold my tongue, or receive unwanted compliments. And what I really wanted, on a Friday night after a long week of work, was just to be understood, and ideally by a handsome, single, thirty-something operator.

Unfortunately for me, and for the terrorists of this world, it looked like the SEALs were out there somewhere in the night working to keep us safe. It was a quiet evening, me alone at the bar, three or four two-tops seated with couples out on dates and Bon Jovi playing quietly on the radio. My phone buzzed on the countertop. I thought it was another group text about the party I wasn't invited to, but I was surprised to see it was Buck, an F/A-18 pilot who had been one of the nicest, most respectful men on deployment. Buck was not only handsome, and an operator of the highest skill, but he was a good man.

I checked the message. I have long since deleted his text, but it read something like, "Hey, haven't seen you in a while. Just checking in. U all good?"

Unlike Chick and Beans, who I felt had inquired into my well-being only to evaluate me and try to improve my performance, I believe Buck was reaching out because he cared about me and wanted to know if his friend who fought alongside him in war was doing okay. Many people knew I was having a tough time. And others were as well. It's natural. And so is checking in. But this was the first such text I'd gotten from a squadronmate—guy or a girl—since being back home from deployment. This gesture, as slight as it was, was like a powerful drug that not only numbed the loneliness and pain that was killing me from the inside out, but also instantly filled me with a profound sense of gratitude and warmth. I turned my body away from the bartender as my eyes welled with tears and I stared at my phone.

"All good," I typed with a thumbs-up emoji.

His response came quickly: "Well, if you ever want to talk about it, LMK. I'm back in town next week."

I watched his blurry response populate in the text window on my iPhone and felt my heartbeat skip in my chest and the hairs on my arm raise. I also watched as those clear lines I had drawn years ago went blurry. Buck had been in my squadron, but had left for his next duty station after we returned from deployment. I had heard he was doing well, professionally, at least. Buck was married, but technically he was either divorced or waiting on the finalization, or so I had heard. I couldn't tell if it was the beer or the prospect of catching up with Buck that was making my head spin, so I left cash for my tab next to my half-drunk Sapporo, and called it a night. I tried to regain my senses as I made my way home, but at the same time I didn't know what to make of Buck's offer. But then as I crossed the threshold and breathed in the loneliness of home, I knew that blurry lines or not, I needed someone to connect with.

★

While there had always been plenty of pent-up sexual tension and attraction between me and Buck, we did not leap into bed. What happened between us slowly developed over long conversations, a little witty banter, lots of listening to each other, and a few too many work trips to the East Coast. To look at him, Buck appeared to be the archetypal pilot, clear-eyed, square-jawed, cut and strong biceps. But as in many cases, looks deceive. While not exactly a navel-gazing poet type, Buck was surprisingly sensitive and a tremendous listener.

In Buck I found a man I could open up to in a way I had not even done with the Minotaur. I shared with him my frustrations, feelings of isolation; I even told him about my suicidal thoughts and the feeling that if I didn't get out of the squadron something bad was going to happen to me. He encouraged me to leave, to try out for a public affairs position with the Blue Angels, which I did. Even though I did not make the team, trying out for the elite organization was one of the highlights of my aviation career. With the Blues I experienced firsthand how respect and professionalism can elevate a group of humans to regularly perform superhuman feats. The team cohesiveness and supportive culture on the Blue Angels was the polar opposite of what I was living with in the Blacklions.

Buck was the only one who seemed to understand why I felt there was something deeply toxic about the Blacklions. He was appalled at how I had been treated by people we had deployed with. I had told him that it was as if the bonds we formed in combat turned from iron to flimsy plastic as soon as we returned to Virginia Beach. Buck, who had been on two deployments with the Blacklions, told me that he understood the group dynamics and saw the shift. Even though no one dared belittle Buck, he did not participate in the bro culture. Thus even when he was in the squadron, he kept to himself.

Since Buck was far away, in a different squadron operating on an unpredictable flight schedule, the two of us were constantly playing phone tag, exchanging short texts throughout the day, and stealing away on trips where we could be alone. We'd meet where work trips took us, whatever we could fit into our busy work schedules.

Buck was a really good pilot, one of the best I've ever flown with, an air-show-quality pilot. We talked about a fantasy life flying performances. I would be Buck's WSO and we would blast around the country in our jet, on the air-show circuit, showing the public what our jet and their tax dollars could do. Crazy as the idea sounds now, we believed it could be within reach, if only for a short time.

Like many illicit relationships, ours was intense and burned quickly. Initially upon reconnecting, I was naive and thought he was happily divorced. Then after finding out he was separated, there was always the idea, at least at first, that he would formalize his separation from his wife in a divorce. He told me one day when we were opening up to one another that he actually hadn't found his own place yet, but because he was gone so much for work that it was like a separation, and when he was home he was sleeping on the couch. Later he claimed he was sleeping on a friend's couch. Finally, I learned, he was at home in his bed all along. The lines I drew were no longer blurred. They were clear and I had crossed well over. Even when I knew this and was aware that I had misunderstood his relationship status and knew there was no chance of Buck leaving his wife, I couldn't get away—despite my best efforts. It was like a black hole sucking me in and I hung on long enough until I felt like I had gotten back some of the strength and confidence I had lost.

Buck helped to heal me, at least partially. And as soon as I was better, I cut him off and ended it for good.

I do not regret my relationship with Buck. He helped pull me up from a dangerously dark place and it would be wrong to regret the good that came out of the relationship, even if it violated my own moral code. That said, I certainly regret any hurt I may have caused his family. I do not envy Buck or his wife. My understanding is that Buck's marriage had been suffering long before me. But that is for him to figure out, not me, and certainly not my place to judge.

An odd by-product of this short affair, which I sincerely hope to be my first and last, was how it changed my view of Navy wives. I now knew firsthand why the wives feel so threatened, why they convert their husbands' flight suits into sexy outfits, why they send centerfold-style photos to their men at sea, and whey they tend to treat women like me with suspicion, resentment, and even outright hostility. Understanding the *why* firsthand and knowing there was a real reason to feel threatened helped me to see myself through their eyes and forgive them.

The women who are married to naval aviators make tremendous sacrifices—quitting successful careers for their husbands' careers, raising children alone, uprooting their families to move around the world, never seeing the fathers of their children more than a few months on end. The wives make those sacrifices so their husbands can risk their lives serving their country.

They also make those sacrifices so their adrenaline-junkie husbands can live out their childhood dreams of having the coolest flying job. Getting to race across the globe at the speed of sound in the most badass vehicles humankind has devised, blowing up bad guys, partying their faces off in every port of call that will take a US supercarrier, and chasing women, even the ones who fly with them. The spouses of active military bear a heavy load for us all. And while I would still like to make a few of the wives eat a fistful of their own hair extensions, most of the feelings of hurt, anger, and resentment I have felt at times toward them have been replaced by a deep and abiding respect.

CHAPTER THIRTY-SEVEN

★

November 2015; NAS Oceana, Virginia Beach, VA

Grammy had just gotten a new iPad and one evening she popped up on FaceTime.

"Grammy?" I answered, not knowing if she'd simply hit the wrong button. I looked at the screen pointed at her ceiling, a fan blade whipping in circles. "Grammy, you there?"

"Oh yes, dear, I'm here. I see your beautiful face."

"Grammy, you have to tilt the camera toward yourself, so I can see you." I watched her ceiling fan whirl round and round.

"That's right, they told me that in my iPad class. How's that? Oh, I see myself in the little screen." She giggled, coming into view. "Well, I was talking to your mother earlier, and she said you might be leaving the squadron . . . and I just thought I'd call . . ."

"I know, it's a bit of a surprise, huh?" I could feel my smile fade. "I'm trying to figure out what's next, if I need to stay flying or move on to something else."

"Honey," Grammy said, looking into my eyes and deep into my soul. "You've done well flying, but if you're not fulfilled, there's no harm in making a change."

"But I've invested so much." I could feel my voice cracking in the back of my throat.

"In the past nine decades, I've learned that there are more important things than just your job. I miss seeing that bright Caroline smile. You did your best, and you're going to succeed in whatever you choose to do. Everyone has to move on at some point; whether you leave the cockpit now or later, in the long run, it doesn't matter."

I heard a knock in the background and saw Grammy's eyes dart toward her front door.

"Earl?" I asked with a smirk.

"Yep," Grammy said. "We got a date."

"Well, don't keep him waiting," I said. "Have fun and call me back soon."

"Love you, dear. Chin up."

The last thing I saw was Grammy's face, three times its normal size as she brought the iPad up close, looking for the hang-up button. I signed us off and thought about her life and what she had just shared with me. When she was my age, Grammy had lived an exciting, adventurous life as a nurse in Ireland during World War II. When she came home from the war, she continued working as a travel nurse, rotating through stints on the Manhattan Project and other top-secret programs at places like the Los Alamos National Laboratory. But eventually she stepped away from her fast-paced, prestigious career to fulfill other facets of her life, and it was then that she met my grandfather. Grammy wasn't a failure because she chose a different path, in fact, it helped her become a better version of herself.

Mulling over what Grammy had said, I also looked at my decision through a lens I could never seem to get away from—aviation. In flight school, one of the most important things we learned was that if your jet is in trouble, plummeting toward terra firma, and the pilot is not able to regain control, the only correct action is to "punch out." Yes, pulling the ejection handle may cost the Navy a very valuable plane, but it preserves the most important thing—human life. Punching out is the last thing you want to do, but it means there's a pretty high likelihood that you will walk away with only cuts and bruises, instead of burning in the wreckage. If you survive, it actually results in mission success.

I'd had plenty of time to think about it, and I knew without a doubt, this plane could not be saved. It was time for me to punch out.

★

I drew a breath, and knocked on Chick's office door.

"What's up, Dutch? Come on in," he said, motioning to a chair in front of his desk.

"Good morning, sir," I said, shutting the door. Beans's office was next door, so I didn't want our voices traveling. I gently placed my notebook in my lap and folded my hands on top. "Sir," I began. "I wanted to talk to you about the way forward," I said, pausing to muster all the courage I had left. "I've made my decision."

"Sounds good, what's it going to be?"

"I've decided that I don't want to fly on my next tour," I said, quickly reciting the words I had practiced over and over.

Chick leaned back and sighed, clearly a little surprised. "Roger that, Dutch. I understand your choice. It's one that I'm sure hasn't been easy to make." He took a moment to think. "Have you considered your options? There something you have in mind?"

Ever the type-A personality, I had come armed with options. "Actually, I came with a list." I opened my notebook to the earmarked page, a smile creeping across my face. All of my dream jobs were in order of preference. "The rotation dates work with my timing. I know the individuals currently holding the jobs, and my qualifications fit the billet requirements. Here," I said, extending the list to the skipper.

"You really did your due diligence. Looks like some great opportunities." He made a couple of notes on my dream sheet and then looked up at me and smiled. "You've done a great job in the squadron. Let me see what I can do."

I closed the door behind me quietly, knowing that my fate was tucked in his hands. In the Navy, as an officer, you cannot apply for any job that you find in the inventory. If you're going for something that is nonstandard, your XO has to go to bat for you and work outside the normal avenues to get you the job. While I knew my XO, Beans, was no help, I was very fortunate that Chick was willing to fight for me. I'd been honest with my boss, now the rest was up to the Navy powers that be. Burden lifted, I let out a deep sigh and hurried down the hallway.

★

Having done everything I could do, I was finally savoring the last bits of life in Virginia Beach. I'd started working out again, running in the early mornings, noticing the tourists stare into the skies as the jets soared above the beach. For once, I was okay being just one of them. Don't get me wrong, my time in the jet was amazing and more fun than can be imagined. Now that my days were numbered, I could fall in love again. Every time I climbed into the cockpit, it was my little secret.

Finally, after waiting on pins and needles, a few weeks later, Chick called me back into his office. "There's something you wanted to talk to me about?" I asked, knowing he was about to reveal my fate.

"Dutch, yes. Got a call this morning and you should be

pleased—you got your first choice. You're going back to the Naval Academy," he said excitedly as he clapped his hands together in Chick-esque fashion.

"Wow, sir! Thank you." Immediately a weight was lifted.

"I'm glad we could get you something that you'll be happy with. But there is one thing I have to ask of you . . . I need you to leave before my change of command," he said.

"Okay," I said slowly, not knowing what he was getting at. It was a strange, out of the blue request, but I was so naive and excited to go back to the Naval Academy I didn't press Chick about it.

Because I was leaving earlier than expected, I only had a short time to get my apartment packed up and moved north. The next month or so was a blur. I wrapped up my flights in the squadron with one final trip to Colorado with one of my favorite pilots, Waldo.

When it was time to fly out, everyone gathered—my parents and friends, even Craig had flown his training plane in. It was a jubilant afternoon; we showed off the jet, and when it came time to launch, the ground crew escorted our entourage in a private bus out to the runway and allowed my family and friends to stand on the side of the runway for our takeoff like they were lining the sidelines in the final quarter of the Super Bowl. Waldo and I taxied onto the runway, and I was shocked that we got our requested clearance.

"Lion 11, cleared for takeoff. Authorized unrestricted climb, flight level two seven zero. Switch Denver Center on 363.1." Like two little kids who'd just been given permission to break all the rules, Waldo and I giggled at the ludicrous request ATC had just approved. Basically they had just given us the clearance to fly our jet Crocket-style, like we stole it.

That detail could wait; for now it was back to business.

Check, check, check. Take-off checks complete. We were

ready to go. I looked out at the crowd gathered down the runway, smiling because I knew they had front-row tickets to something so intense, something they'd never experienced before.

Waldo revved the engines and released the brakes.

"Off the peg," I said over ICS, letting him know that the airspeed was alive. I felt the afterburners click into MAX. "Hundred knots," I said, keeping with my normal procedures. The airspeed indicator was ticking up, and I knew we would be airborne momentarily. We accelerated through our rotation speed, and Waldo barely lifted the nose of the jet so we were hovering a mere ten to twenty feet above the ground as we screamed by our spectators. All was normal as the gear clicked in place, but instead of gently climbing away from the runway, Waldo held the jet steady at twenty feet. Our airspeed continued to build and we ripped down the eleven-thousand-foot runway faster than a drag racer at the Outlaw Drag Racing Championship. Just before we ran out of pavement below us, Waldo stood the jet on its tail and we rocketed skyward like we were headed for the moon.

I couldn't contain my excitement any longer. I belted out with laughter, "Dude, that was incredible!"

He giggled like a little kid who got caught coloring on his parents' freshly painted walls. "There was a flock of birds at the end of the runway and I was just trying to avoid them. Safety first."

I wasn't expecting that kind of send-off and neither were any of the spectators lining the runway, but it was the cherry on top of my last F/A-18 flight.

★

After the movers packed up my apartment, my last stop before heading out of Virginia Beach was to the squadron to pick up my paperwork and flight gear. I'd come to find out the reason for the skipper's strange request. He was having me leave before his change of command so that he could save my ranking spot for one of the lower-ranked JOs who'd be staying in the community. Leaving early meant I wouldn't get my deserved high-water FITREP, or evaluation report, which meant that I'd never promote past lieutenant in the Navy, at least not in the fighter community.

This is akin to a fourth-year surgery resident requesting a transfer from one hospital to another because she was unhappy and suffering under relentless harassment, and then learning that her supervisor would grant her transfer, but as part of the bargain she'd forever remain a resident, would never be able to become a surgeon and her career and ability to rise in her field would be all but ruined. After all the work, the sacrifice, the commitment, and love of the fighter community, this was a final knife in my back. And the worst part is in that moment, I didn't care. I knew I'd hurt later, but when the realization hit me I didn't fight it, I didn't even feel it. I said nothing. Didn't even flinch. In hindsight, years later I processed what I believe to have been Chick's intended outcome; he wanted to protect the best rankings for those who would remain in his beloved fighter community, but in doing so he actually ensured the fifty million dollars the Navy had invested in me was walking out the door.

I gathered my things, said my goodbyes, shook Chick's hand, and walked out of the squadron one last time, holding my head high as I exited the flightline gate and walked one foot in front of the other to my Cayenne parked outside.

I slid into the driver's seat, and with the door still open, I unlocked my phone and deleted myself from the JO text message group. That was it. I was no longer a Jet Girl, not to them anyway. I paused to take a deep, cathartic breath.

I closed the door, pulled the seatbelt across my chest. I turned the ignition. The Porsche engine roared to life, and I felt the familiar rumble in my chest. *Check.* I shifted into first gear, glancing into my mirrors. Nothing behind me. I turned my eyes to the road ahead. Clear. *Check, check.* I hit the gas. The gold SUV lunged forward, increasing speed as I took her up to third gear. I was only going thirty, but my heart raced ahead.

Finally, I was free.

EPILOGUE

★

On a cold February afternoon, I hustled across the campus of the United States Naval Academy in my flight suit, my worn leather aviator jacket around my shoulders, blocking the brisk breeze blowing off the Severn. Since leaving the fighter community I had come back to USNA with a very clear set of goals. I was there to teach leadership, develop the next generation of naval officers, recruit aviators, revamp the system, and simply make it better.

To accomplish these goals, I had to use a number of strategies and tactics. First, I had my teaching job, which allowed me to imbue the next generation with lessons about leadership and course-correct the system that way; however, we all know classwork is often a means to an end—usually a passing grade and some wisdom that can be used, but unless put into practice will inevitably fade from consciousness. That wouldn't do. I not only wanted more but I'd been sent to the Academy to bring about real change. Senior leadership had given me a new mission: to increase the recruitment and retention of the best candidates for naval aviation, in particular women.

For the past few years, the top graduates were service selecting into cyber, intel, SEALs, and even SWO instead of naval aviation, whereas historically, aviation attracted the best and brightest. We were having to force students into the business.

That night I was hosting a dinner for 315 pilot and NFO selectees (Naval Academy seniors selected to fly), USNA staff aviators, and twenty-five aviation admirals from the fleet. All the biggest bigwigs in aviation would be there. The purpose: to welcome the young midshipmen selectees to the community and motivate them for their exciting career ahead. But there was far more than that at stake.

Given the volatile power struggle unfolding between the West and our adversaries in Russia, China, North Korea, and Iran, the need for superior naval aviators projecting power at the tip of the spear has never been more critical. I witnessed firsthand the malice of the Taliban and the savagery of ISIS. While these threats were very real and counter to our principles, the threat posed by those enemies is mere child's play when compared to the mounting tensions in the Far East. Every day, our adversaries strengthen their military forces and challenge America's national security strategy. Though we aren't currently engaged in direct conflict with those nation states, should the day come we must be ready, or the United States and our allies will pay the ultimate price.

Preparing for this technologically advanced and tactically complex battlefield requires tireless work to maintain our edge. One of the primary reasons, if not the single greatest reason, that the US remains the premier world power is the superiority of our warfighters. To put it simply, our national security and personal safety depend on the armed forces recruiting top men and women, selecting the best for the hardest roles and then preparing them for battle.

This is why the US Navy invests so much in aviators, especially those operating in the most pivotal combat roles. Of the class of 315 selectees, only 20 percent will make it into the jet community, and for those, US taxpayers will invest upwards of $6 million per student in initial training costs. On top of the training costs, there is an *enormous* amount spent on aviator flying experience. On average, a first-tour junior officer will depart their fleet squadron with almost one thousand hours in an F/A-18. Each flight hour costs the Navy $50,000 in maintenance and fuel costs, meaning that each aviator's flight time alone is worth $50 million. Fifty million dollars! So it's not only important to recruit and train the best, it's vitally important to retain them. And this applies to both men and women.

And yet the fighter community has been experiencing unusually high junior-officer attrition rates. My own experience is just one instance. The data confirms this. While part of the trend can be attributed to surging airline hires and an improving economy, those factors don't explain everything. The numbers speak for themselves: During my time in the cockpit, 52 percent of male F/A-18 aviators left at their first opportunity, and a staggering 83 percent (four out of every five) of women in the fighter community got out at their first chance. Half of the men are leaving, but women are going at a precipitously higher rate. Thirty-one percent higher, to be exact. At the time I was in the cockpit, women made up only 1.7 percent of the fighter community*; with such high attrition rates this means there are very few women promoting to senior ranks in naval aviation. This is why there was more at stake at that February dinner than just welcoming young, aspiring members of a community. It was the first step in a long, challenging, expensive journey that the Navy takes with its warriors.

The evening was the standard Navy affair, draped in pomp and circumstance and steeped in tradition. It began with a cocktail hour in a cavernous, beautiful hall with twinkling lights cre-

---

* Navy Retention Monitoring System. *Strength Counts of Active Component Aviators with F/A-18 Additional Qualification Codes (AQD)*. July 2014. Raw data.

ating a whimsically festive ambiance. As I strode in from the cold and took it all in, a glimpse of white material punctuated by rich, caramel-hued wood suspended from the ceiling caught my eye. No, it wasn't typical ballroom decor, it was a life-sized replica of the first plane ever purchased and flown by the US Navy, the Wright Brothers' Model B-1, complete with pontoons and all of its original specs from 1911. I watched as the midshipmen took in the sight, a reminder that in five short generations, we have progressed from rudimentary flying machines of wood, leather, and fabric to nuclear-powered aircraft carriers and jet engines.

After dinner was served and rounds of speeches given, the night wrapped up and the real work began—at the bar where the pilot and NFO selectees crowded around me and peppered me with questions.

"Dutch, what's it like? Flying in-country?"

I told them about mountain peaks in Afghanistan, conducting shows of force to suppress the enemy, and I even cracked them up with my story of peeing in the cockpit. Of course they were interested, but I could tell they were only half listening, their far-off looks giving away that they were envisioning themselves in the jet, years in the future. They were dreamy-eyed and excited, and as they loosened up, I knew I was doing my job—enchanting them with the very life I loved so much.

A few drinks in, Airboss, the three-star admiral in charge of naval aviation, strolled over and broke up the party. "Whoa, whoa!" he joked to the midshipmen. "Can any of you get a word in edgewise over here with Dutch holding court? What kind of lies is she telling you guys!"

We all laughed and finished our rounds. Like modern-day Cinderellas, the midshipmen's liberty expired at midnight, and they rushed to get back to the Academy in time.

Airboss, his aide, and I were the only ones left. "How 'bout a nightcap, Dutch?" Airboss asked.

"Of course."

Airboss ordered a round of Basil Hayden bourbon and we saddled up to the bar. He angled his stool toward me, took a sip, and I felt the evening take a serious turn.

"Dutch . . ." he began. "I've been looking at data from the Academy over the past two years and since you arrived at USNA, I want you to know . . . you've turned the numbers around."

"Thank you, sir." I nodded, but he went on.

"It's hard to believe that for a while the Academy was forcing midshipmen into aviation. Now, for the first time in recent history, we selected all first-choice pilots and NFOs. You and your team transformed the aviation culture and poured your heart and soul into getting the best and brightest into the cockpit. It's had a huge impact on the midshipmen and the naval aviation enterprise. Really, Dutch, you've done a phenomenal job."

There was an awkward pause. I was speechless. I couldn't believe that after all I had been through I was being praised by the admiral in charge of all naval aviation. "Thank you, sir, but I was just doing my job. Trying to open midshipmen's eyes to what I believe is the best career in the world."

"Flying *is* the best job in the world." He grinned. "So what's next? What job do you want? Do you want to go back and fly?"

This took me aback. I hadn't been in a jet in eighteen months. Leaving the Blacklions and the cockpit for the Academy, I'd always believed my flying career was over. "Sir, don't toy with my emotions, you know how much I love it."

"Well, why not? You've done an amazing job here, tackling our recruiting challenges. Your hard work should be rewarded. I wanna know what we can do to keep you . . ." He trailed off, rubbing his forehead as if getting rid of a headache, and then started again. "To be honest, what hurts my head, what keeps me up at night is that aviators—especially women like you—are leaving the jet community." He took another long sip of his bourbon. "And I can't figure it out. Why, Dutch, if this is the best job in the world, are we losing our top aviators?"

Here we go, I thought. He's getting to it.

"Sir, that's the million-dollar question, isn't it?" I began hesitantly. "I don't think we're losing our best because of the job. It's not the long deployments, or even the bodily abuse from the cat shots or traps." I paused, wondering how honest I should be. "I know I'm not speaking for everyone, but my hesitations with getting back into the cockpit revolve around a few things . . . the issues are more comprehensive than we have time for tonight, but here goes. I would love to take your offer to go back to a flying job, but while I could survive in the F/A-18 community, I don't think I will ever thrive. Systemically, women are still treated and evaluated vastly differently than our male counterparts. I know it's 2017, and we think those issues are behind us, but based on my experiences and those of my peers, and much of the research I've been doing, we have a long ways to go."

"What, specifically?" He set down his drink.

"Well, for example, there are bad apples in the community that perpetuate harassment and gender discrimination, and while most of their behaviors are not outright criminal, over time, they wear you down. I have been on the receiving end of that for too long." I couldn't believe I said it.

His eyes narrowed.

"It's not everyone, and I can only speak for myself, sir. But in my view, the fighter aviation culture accepts this behavior as a norm. And since women still make up such a small percentage of the F/A-18 population, our voices aren't heard, we're not promoting and rising up, so we don't have many like-gendered role models and mentors at the top . . . and to be frank, sir, we don't experience many of the same perks of the community that the men do."

"We try to make a fair system . . . but . . ." He nodded slightly. "I think you're right."

I wanted to say, *there's no thinking about it.* After all, there's hard data to back this up. "Sir, I've really been researching this. I'm sure you know the Navy conducted a confidential study in 1996 on gender and racial integration in naval aviation, and if you look at the results, you have your answers." My voice grew heavy as I thought back. "Every single negative incident or slight I experienced in the Blacklions was listed in a behaviors chart from back in the nineties, proving that as far as things have come in the past few decades, not much has changed. Basically, if you look at it, these problems occurred to more than 63 percent of women in naval aviation back then, and now you're losing four out of every five women at their first opportunity. There *has* to be a correlation."

1996 NEOSH survey gender discrimination behaviors experienced during the past 12 months (percentage "Yes")

	MALE OFFICERS		FEMALE OFFICERS	
	AVIATION	REST OF NAVY	AVIATION	REST OF NAVY
Negative Comments	1	5	63	31*
Offensive Jokes	1	3	39	23
Ignored by Others	1	3	39	20*
Given Menial Jobs	0	2	28	9*
Not Asked to Socialize	0	4	37	15*
Denied Potential Reward/Benefit	0	2	4	5
Physically Threatened	0	1	3	3
Physically Assaulted	0	0	4	1

*Statistically significant difference ($p < .01$) between Aviation Squadron and Rest of Navy.

From Uriell, Zannette A., and Paul Rosenfeld. "Minorities and Women in Naval Aviation Training: A Look Back at a 1997 Study." Navy Personnel Research, Studies, and Technology, NPRST-TN-11-2, Millington, TN (November 2010).

"Do you have the study with you?"

I drew out my phone and held it out. Airboss and his aide leaned in. I wondered now if the admiral had bargained on this when he asked for me to have a simple nightcap with him.

"This is still going on, isn't it?" he said.

Part of me wanted to laugh or sigh or scream. "Sir, it's a struggle everywhere, in the military and in the private sector. But what I care about is *our* community, the fighter community." I cleared my throat. "And, in my experience, that community refuses to openly address the cultural issues that are driving their minorities out the door. Look, women are using their feet to speak to leaders. They're walking out the door. Women can tell when they're not truly wanted in an organization, so why would they stay?"

"I see . . ." he said. "You've made great points, Dutch. Clearly you've thought long and hard about this. So how do we fix it?"

This is what I was waiting for.

"In my opinion, it's fairly simple . . . we need to change the culture. We have to get back to the core principles of naval aviation—fly, fight, lead, and win. We must do that today and . . ." I trailed off, carefully considering my next words. "Not sure if I should speak so freely."

"And?" he said. "Tell me."

"And it starts from the top. From Airboss down to the most junior ranks, we need to hold ourselves to a higher standard today, now. So that tomorrow, we are better. More battle-ready."

"What does that look like?" He glanced at his aide, whose eyes seemed to say, *listen to her.*

I had the floor. Again, I weighed my words carefully. "As a community, we need to do better. We need to *be* better. We need to treat one another with respect—not just the white guys, but the women and minorities need to be treated equally, made to feel like they belong, and given the same opportunities. Leadership must advocate for its women; mentoring, sponsoring, and grooming them like they would their white male counterparts. It's about recognizing that we all have unconscious biases and overcoming that by advocating for all types. Not only that, but all aviators, regardless of race, gender, or background, need to be objectively evaluated on their job performance, rather than promoting only those who fit the stereotypical mold of what a fighter pilot traditionally used to be. The jet doesn't know or care who is in the cockpit. The jet only cares that people in the cockpit perform at the highest level."

Airboss, himself a seasoned pilot, knew what I was talking about. "It's true." He nodded. "When you're in a jet, the best operator wins, period."

"Yes, sir," I said, draining the last of my drink. "And when we all believe that, the problem will be solved."

ACKNOWLEDGMENTS

★

Working on this book took over two years and a village. I'd like to thank those who helped bring it to life.

Tod Williams: *Jet Girl* was your brainchild. Your enthusiasm, patience, empathy, and the understanding of your family helped me take this manuscript from an idea to publishing in a way that transcends the military/civilian divide.

Nancy Johnson: Mom, always my #1 supporter, you believed in *Jet Girl* and stood in superbly when my plate was full.

Elizabeth Bolke: A total lifesaver and magician. You swept in and helped transform our manuscript into beautiful prose, all while raising three amazing littles.

Elizabeth Beier and your incredible team at St. Martin's: Your passion and vision for the manuscript and everlasting patience helped more than you can know.

Ian Kleinert: My agent, and man with the plan.

Guy Bevan: Your artistic vision and graphic support brought my dream cover to life. Thank you for your persistence and collaboration (even from fourteen time zones away).

★

As *Jet Girl* is based on my life, I owe the deepest debt of gratitude to those who made an indelible mark on me and onto these pages.

To my family: Hazel Johnson—Grammy, after a century on this planet you are the matriarch, a huge influence, and my spirit animal. Grampy (RIP), Marty, Craig, Lauren, Elin, and my pup Lily—you are my biggest supporters, especially Dad, who has always worked hard to turn our dreams into a reality and starts conversations with, "Let me tell you about my son and daughter who fly Navy jets."

Sharktank, the Jet Girls, and the badass women of the skies: Carolyn, Mere, Aly, Kelsey, Gosho, Ash, Hadley, and all the others—you are beautiful, intensely talented, and powerful women who helped me navigate life and war.

My fellow aviators, mentors, and allies: Thank you for accepting the authentic version of me—bright spots, flaws, and all.

Gizzard & VA, Minotaur, Ang & Dave, Gringa, Rubyn, FASM, COTL, Swampy, Waldo, Peeper, Buzz, #winner, Crocket, BoomBoom, Fatty, Bullet, MM, Andrew L., Kone, R10, Doug, Kelly, Lisa & Chris, Troy & Suzanne, Jaime, the Brewer Babes, Junie (RIP), Teri, Karen (RIP), Eileen, and Kim the SEAL mom.

To my Torch Lake family: The Palazzos, the Caulkins, the Burkhams, the Norrises, and to the many others not listed, you are fine Americans, thank you for your friendship and all you've done for me.